William Roscoe Thayer

Throne-Makers

William Roscoe Thayer

Throne-Makers

ISBN/EAN: 9783337258474

Printed in Europe, USA, Canada, Australia, Japan

Cover: Foto ©berggeist007 / pixelio.de

More available books at **www.hansebooks.com**

THRONE-MAKERS

BY

WILLIAM ROSCOE THAYER

BOSTON AND NEW YORK
HOUGHTON MIFFLIN COMPANY
The Riverside Press Cambridge

TO
DR. MORRIS LONGSTRETH
IN MEDICINE, ORIGINAL AND WISE

IN FRIENDSHIP, STEADFAST

PREFACE

Since 1789 every European people has been busy making a throne, or seat of government and authority, from which its ruler might preside. These thrones have been of many patterns, to correspond to the diversity in tastes of races, parties, and times. Often, the business of destroying seems to have left no leisure for building. In England alone have men learned how to remodel a throne without disturbing its occupant; as we in America raise or move large houses without interrupting the daily life of the families who dwell in them.

To portray the personality of some of the conspicuous Throne-Makers of the century is the purpose of the following studies. I have wished to show just enough of the condition of the countries under review to enable the reader to understand what Bismarck, or Napoleon III, or Kossuth, or Garibaldi, achieved. I have been brief, and yet I trust that this method has afforded scope for exhibiting that influence of the individual on the multitude which — however our partial science may try to belittle it — was never more strikingly illustrated than by such careers as these in our own time.

The group of Portraits which follow require no special introduction. In the "Tintoret" and "Giordano Bruno" I have brought together as compactly as possible, for the convenience of English readers, what little is known about these two men. Berti's work on Bruno, from which I have drawn largely, deserves a wider recognition than it has received outside of Italy; whoever reads it will regret that that eminent scholar was prevented from completing his volume on Bruno's philosophy. The sketch of Bryant was written in 1894, that of Carlyle in 1895, on the occasion of their centenaries.

My thanks are due to the proprietors of *The Atlantic Monthly*, *The Forum*, and *The American Review of Reviews* for permission to reprint such of the following articles as originally appeared in those periodicals.

W. R. T.

8 BERKELEY STREET, CAMBRIDGE,
December 8, 1898.

CONTENTS

THRONE-MAKERS: PAGE

    BISMARCK 3
    NAPOLEON III 44
    KOSSUTH 79
    GARIBALDI 115

PORTRAITS:

    CARLYLE 163
    TINTORET 193
    GIORDANO BRUNO 252
    BRYANT 309

THRONE-MAKERS

BISMARCK

One by one the nations of the world come to their own, have free play for their faculties, express themselves, and eventually pass onward into silence. Our age has beheld the elevation of Prussia. Well may we ask, " What has been her message? What the path by which she climbed into preëminence? " That she would reach the summit, the work of Frederick the Great in the last century, and of Stein at the beginning of this, portended. It has been Bismarck's mission to amplify and complete their task. Through him Prussia has come to her own. What, then, does she express?

The Prussians have excelled even the Romans in the art of turning men into machines. Set a Yankee down before a heap of coal and another of iron, and he will not rest until he has changed them into an implement to save the labor of many hands; the Prussian takes flesh and blood, and the will-power latent therein, and converts them into a machine. Such soldiers, such government clerks, such administrators, have never been manufactured elsewhere. Methodical, punctilious, thorough, are those officers and officials. The government which makes them relies not on sudden spurts, but on the cumulative force of habit. It substitutes rule for whim; it suppresses individual spontaneity, unless this can be transformed into energy for the great machine to use. That Prussian system takes a turnip-fed peasant, and in a few months makes of him a military weapon, the length of whose stride is prescribed in centimetres — a machine which presents arms to a passing lieutenant with as much gravity and precision as if the fate of Prussia hinged on that special act. It takes the average tradesman's son, puts him into the educational mill, and brings him out a professor, — equipped even to the spectacles, — a nonpareil of knowledge, who fastens on some subject, great or small, timely or remote, with the dispassionate persistence of a leech; and who, after many years, revolutionizes our theory of Greek roots, or of microbes, or of religion. Patient and noiseless as the earthworm, this scholar accomplishes a similarly incalculable work.

A spirit of obedience, which on its upper side passes into deference not always distinguishable from servility, and on its lower side is not always free from arrogance, lies at the bottom of the Prussian nature. Except in India, caste has nowhere had more power. The Prussian does not chafe at social inequality, but he cannot endure social uncertainty; he must know where he stands, if it be only on the bootblack's level. The satisfaction he gets from requiring from those below him every scrape and nod of deference proper to his position more than compensates him for the deference he must pay to those above him. Classification is carried to the fraction of an inch. Everybody, be he privy councilor or chimney-sweep, is known by his office. On a hotel register you will see such entries as "Frau X, widow of a school-inspector," or "Fräulein Y, niece of an apothecary."

This excessive particularization, which amuses foreigners, enables the Prussian to lift his hat at the height appropriate to the position occupied by each person whom he salutes. It naturally develops acuteness in detecting social grades, and a solicitude to show the proper degree of respect to superiors and to expect as much from inferiors, — a solicitude which a stranger might mistake for servility or arrogance, according as he looked up or down. Yet, amid a punctilio so stringent, fine-breeding — the true politeness which we associate with the word "gentleman" — rarely exists; for a gentleman cannot be made by the rank he holds, which is external, but only by qualities within himself.

Nevertheless, these Prussians — so unsympathetic and rude compared with their kinsmen in the south and along the Rhine, not to speak of races more amiable still — kept down to our own time a strength and tenacity of character that intercourse with Western Europeans scarcely affected. Frederick the Great tried to graft on them the polished arts and the grace of the French: he might as well have decorated the granite faces of his fortresses with dainty Parisian wall-paper. But when he touched the dominant chord of his race, — its aptitude for system, — he had a large response. The genuine Prussian nature embodied itself in the army, in the bureaucracy, in state education, through all of which its astonishing talent for rules found congenial exercise. One dissipation, indeed, the Prussians allowed themselves, earlier in this century, — they reveled in Hegelianism. But even here they were true to their instinct; for the philosophy of Hegel commended itself to them because it assumed to reduce the universe to a system, and to pigeon-hole God himself.

We see, then, the elements out of which Prussia grew to be a strong state, not yet large in population, but compact and carefully organized. Let us look now at Germany, of which she formed a part.

We are struck at once by the fact that until 1871 Germany had no political unity. During the centuries when France, England, and Spain were being welded into political units by their respective dynasties, the great Teutonic race in Central Europe escaped the unifying process. The Holy Roman Empire — at best a reminiscence — was too weak to prevent the rise of many petty princedoms and duchies and of a few large states, whose rulers were hereditary, whereas the emperor was elective. Thus particularism — what we might call states' rights — flourished, to the detriment of national union. At the end of the last century, Germany had four hundred independent sovereigns: the most powerful being the King of Prussia; the weakest, some knight whose realm embraced but a few hundred acres, or some free city whose jurisdiction was bounded by its walls. When Napoleon, the great simplifier, reduced the number of little German states, he had no idea of encouraging the formation of a strong, coherent German Empire. To guard against this, which might menace the supremacy of France, he created the kingdoms of Bavaria and Westphalia, and set up the Confederation of the Rhine. After his downfall the German Confederation was organized, — a weak institution, consisting of thirty-nine members, whose common affairs were regulated by a Diet which sat at Frankfort. Representation in this Diet was so unequal that Austria and Prussia, with forty-two million inhabitants, had only one eighth of the votes, while the small states, with but twelve million inhabitants, had seven eighths. Four tiny principalities, with two hundred and fifty thousand inhabitants each, could exactly offset Prussia with eight millions. By a similar anomaly, Nevada and New York have an equal representation in the United States Senate.

From 1816 to 1848 Austria ruled the Diet. Yet Austria was herself an interloper in any combination of German states, for her German subjects, through whom she gained admission to the Diet, numbered only four millions; but her prestige was augmented by the backing of her thirty million non-German subjects besides. Prussia fretted at this Austrian supremacy, fretted, and could not counteract it. Beside the Confederation, which so loosely bound the German particularists together, there was a Customs Union, which, though simply commercial, fostered among the Germans the idea of common interests. The spirit of nationality, potent everywhere, awakened also in the Germans a vision of political unity, but for the most part those who beheld the vision were unpractical; the men of action, the rulers, opposed a scheme which enfolded among its possibilities the curtailing of their autocracy through the adoption of constitutional government. No state held more rigidly than Prussia the tenets of absolutism.

Great, therefore, was the general surprise, and among Liberals the joy, at the announcement, in February, 1847, that the King of Prussia had consented to the creation of a Prussian Parliament. He granted to it hardly more power than would suffice for it to assemble and adjourn; but even this, to the Liberals thirsty for a constitution, was as the first premonitory raindrops after a long drought. Among the members of this Parliament, or Diet, was a tall, slim, blond-bearded, massive-headed Brandenburger, thirty-two years old, who sat as proxy for a country gentleman. A few of his colleagues recognized him as Otto von Bismarck; the majority had never heard of him.

Bismarck was born at Schönhausen, Prussia, April 1, 1815. His paternal ancestors had been soldiers back to the time when they helped to defend the Brandenburg March against the inroads of Slav barbarians. His mother was the daughter of an employee in Frederick the Great's War Office. Thus, on both sides his roots were struck in true Prussian soil. At the age of six he was placed in a Berlin boarding-school, of which he afterward ridiculed the "spurious Spartanism;" at twelve he entered a gymnasium, where for five years he pursued the usual course of studies, — an average scholar, but already noteworthy for his fine physique; at seventeen he went up to the University at Göttingen. In the life of a Prussian, there is but one period between the cradle and the grave during which he escapes the restraints of iron-grooved routine: that period comprises the years he spends at the university. There a strange license is accorded him. By day he swaggers through the streets, leering at the women and affronting the men; by night he carouses. And from time to time he varies the monotony of drinking-bouts by a duel. Such, at least, was the life of the university student in Bismarck's time. At Göttingen, and subsequently at Berlin, he had the reputation of being the greatest beer-drinker and the fiercest fighter; yet he must also have studied somewhat, for in due time he received his degree in law, and became official reporter in one of the Berlin courts. Then he served as referendary at Aix-la-Chapelle, and passed a year in military service.

At twenty-four he set about recuperating the family fortunes, which had suffered through his father's incompetence. He took charge of the estates, devoted himself to agriculture, and was known for many miles round as the "mad squire." Tales of his revels at his country house, of his wild pranks and practical jokes, horrified the neighborhood. Yet here, again, his recklessness did not preclude good results. He made the lands pay, and he tamed into usefulness that restive animal, his body, which was to serve as mount for his mighty soul. Some biographers, referring to his bucolic apprenticeship, have compared him to Cromwell; in his youthful roistering he reminds us of Mirabeau.

To the Diet of 1847 the mad squire came, and during several sittings he held his peace. At last, however, when a Liberal deputy declared that Prussia had risen in arms in 1813, in the hope of getting a constitution quite as much as of expelling the French, the blond Brandenburger got leave to speak. In a voice which seemed incongruously small for his stature, but which carried far and produced the effect of being the utterance of an inflexible will, he deprecated the assertions just made, and declared that the desire to shake off foreign tyranny was a sufficient motive for the uprising in 1813. These words set the House in confusion. Liberal deputies hissed and shouted so that Bismarck could not go on; but, nothing daunted, he took a newspaper out of his pocket and read it, there in the tribune, till order was restored. Then, having added that whoever deemed that motive inadequate held Prussia's honor cheap, he strode haughtily to his seat, amid renewed jeers and clamor. Such was Bismarck's parliamentary baptism of fire.

Before the session adjourned, the deputies had come to know him well. They discovered that the mad squire, the blunt "captain of the dikes," was doubly redoubtable; he had strong opinions, and utter fearlessness in proclaiming them.

His political creed was short, — it comprised but two clauses: "I believe in the supremacy of Prussia, and in absolute monarchy." More royalist than the King, he opposed every concession which might diminish by a hair's breadth the royal prerogative. Constitutional government, popular representation, whatever Liberals had been struggling and dying for since 1789, he detested. Democracy, and especially German democracy, he scoffed at. For sixty years reformers had been railing at the absurdities of the Old Régime; they had denounced the injustice of the privileged classes; they had made odious the tyranny of paternalism. Bismarck entered the lists as the champion of "divine right," and first proved his strength by exposing the defects of democracy.

Those who believe most firmly in democracy acknowledge, nevertheless, that it has many objections, both in theory and in practice. Universal suffrage — the abandoning of the state to the caprice of millions of voters, among whom the proportion of intelligence to ignorance is as one to ten — seems a process worthy of Bedlam. The ballot-box is hardly more accurate than the dice-box, as a test of the fitness of candidates. Popular government means party government, and parties are dogmatic, overbearing, insincere, and corrupt. The men who legislate and administer, chosen by this method, avowedly serve their party, and not the state; and though, by chance, they should be both skilful and honest, they may be overturned by a sudden revulsion of the popular will. Such a system breeds a class of professional politicians, — men who make a business of getting into office, and whose only recommendation is their proficiency in the art of cajoling voters. A government should be managed as a great business corporation is managed: it has to deal with the weightiest problems of finance, and with delicate diplomatic questions, for which the trained efforts of judicious experts are needed; but instead of being intrusted to them, it is given over to politicians elected by multitudes who cannot even conduct their private business successfully, much less entertain large and patriotic views of the common welfare. To decide an election by a show of hands seems not a whit less absurd than to decide it by the aggregate weight or the color of the hair of the voters. We speak of the will of the majority as if it were infallibly right. The vast majority of men to-day would vote that the sun revolves round the earth: should this belief of a million ignoramuses countervail the knowledge of one astronomer? Shall knowledge be the test of fitness in all concerns except government, the most critical, the most far-reaching and responsible of all? Majority rule substitutes mere numbers, bulk, and quantity for quality. Putting a saddle on Intelligence, it bids Ignorance mount and ride whither it will, — even to the devil. It is the dupe of its own folly; for the politicians whom it chooses turn out to be, not the representatives of the people, but the attorneys of some mill or mine or railway.

These and similar objections to democracy Bismarck urged with a sarcasm and directness hitherto unknown in German politics. When half the world was repeating the words "Liberalism," "Constitution," "Equality," — as if the words themselves possessed magic to regenerate society, — he insisted that firm nations must be based upon facts, not phrases. He had the twofold advantage of invariably separating the actual from the apparent, and of being opposed by the most incompetent Liberals in Europe. However noble the ideals of the German reformers, the men themselves were singularly incapable of dealing with realities. Nor should this surprise us; for they had but recently broken away from the machine we have described, and as they had not yet a new machine to work in, they whirled to and fro in vehement confusion, the very rigidity of their previous restraint increasing their dogmatism and their discord.

The revolution of 1848 soon put them to the ordeal. The German Liberals aimed at national unity under a constitution. Like their brothers in Austria and Italy, they enjoyed a temporary triumph; but they could not construct. Their Parliament became a cave of the winds. Their schemes clashed. By the beginning of 1850 the old order was restored.

During this stormy crisis, Bismarck, as deputy in two successive Diets, had resolutely withstood the popular tide. He regarded the revolutionists as men in whom the qualities of knave, fool, and maniac alternately ruled; the revolution itself, he said, had no other motive than "a lust of theft." One of its leaders he dismissed as a "phrase-watering-pot." The right of assemblages he ridiculed as furnishing democracy with bellows; a free press he stigmatized as a blood-poisoner. When the imperial crown was offered to the King of Prussia, Bismarck argued against accepting it; he would not see his King degraded to the level of a mere "paper president."

Such opposition would have made the speaker conspicuous, if only for its audacity. His enemies had learned, however, that it required a strong character to support that audacity continuously. They tried to silence him with abuse; but their abuse, like tar, added fuel to his fire. They tried ridicule; but their ridicule had too much of the German dulness to wound him. They called him a bigoted Junker, or squire. "Remember," he retorted, "that the names Whig and Tory were first used opprobriously, and be assured that we will yet bring the name Junker into respect and honor." Many anecdotes are told illustrating his quick repulse of intended insult or his disregard of formality. He was not unwilling that his enemies should remember that he held his superior physical strength in reserve, if his arguments failed. Yet on a hunting-party, or at a dinner, or in familiar conversation, he was the best of companions. Germany has not produced another, unless it were Goethe, so variedly entertaining; and Goethe had no trace of one of Bismarck's characteristics, — humor. He possessed also tact and a sort of Homeric geniality which, coupled with unbending tenacity, fitted him to succeed as a diplomatist.

In 1851 the King appointed him to represent Prussia at the German Diet, which sat at Frankfort. The outlook was gloomy. Prussia had quelled the revolution, but she had lost prestige. Unable to break asunder the German Confederation or to dominate it, she had signed, at Olmütz, in the previous autumn, a compact which acknowledged the supremacy of her old rival, Austria. While the humiliation still rankled, Bismarck entered upon his career. Hitherto not unfriendly to Austria, because he had looked upon her as the extinguisher of the revolution, which he hated most of all, he began, now that the danger was over, to give a free rein to his jealousy of his country's hereditary competitor. In the Diet, the Austrian representative presided, the rulings were always in Austria's favor, the majority of the smaller states allowed Austria to guide them. Bismarck at once showed his colleagues that humility was not his rôle. Finding that the Austrian president alone smoked at the sittings, he took out his own cigar and lighted it, — a trifle, but significant. He resisted every encroachment, and demanded the strictest observance of the letter of the law. Gradually he extended Prussia's influence among the confederates. He unmasked Austria's insincerity; he showed how honestly Prussia walked in the path of legality; until he slowly created the impression that wickedness was to be expected from one, and virtue from the other.

During seven years Bismarck held this outpost, winning no outward victory, but storing a vast amount of knowledge about all the states of the Confederation, their rulers and public men, which was subsequently invaluable to him. His dispatches to the Prussian Secretary of State, his reports to the King, form a body of diplomatic correspondence unmatched in fulness, vigor, directness, and insight. With him, there was no ambiguity, no diplomatic circumlocution, no German prolixity. He sketched in indelible outlines the portraits, corporal or mental, of his colleagues. He criticised the policy of Prussia with a brusqueness which must have startled his superior. He reviewed at longer range the political tendencies of Europe. Officially, he kept strictly within the limits of his instructions; but his own personality represented more than he could yet officially declare, — Prussia's ambition to become the leader of Germany. In all his dispatches, and in all places where caution did not prescribe silence, he reiterated his Cato warning, "Austria must be ousted from Germany."

Do not suppose, however, that Bismarck's political greatness was then discerned. Probably, had you inquired of Germans forty years ago, "Who among you is the coming statesman?" not one would have replied, "Bismarck." At the opera, we cannot mistake the hero, because the moonlight obligingly follows him over the stage; in real life, the hero passes for the most part unrecognized, until his appointed hour; but the historian's duty is to show how the heroic qualities were indubitably latent in him long before the world perceived them.

In 1859 Bismarck was appointed ambassador at St. Petersburg, where he stayed three years, when he was transferred to Paris. This completed his apprenticeship, for in September, 1862, he was recalled to Berlin to be minister-president.

His promotion had long been mooted. The new King William — a practical, rigid monarch, with no Liberal visions, no desire to please everybody — had been for eighteen months in conflict with his Parliament. He had determined to reorganize the Prussian army; the Liberals insisted that, as Parliament was expected to vote appropriations, it should know how they were spent. William at last turned to Bismarck to help him subjugate the unruly deputies, and Bismarck, with a true vassal's loyalty, declared his readiness to serve as "lid to the saucepan." Very soon the Liberals began to compare him with Strafford, and the

King with Charles I, but neither of them quailed. "Death on the scaffold, under certain circumstances, is as honorable," Bismarck said, "as death on the battlefield. I can imagine worse modes of death than the axe." Hitherto he had strenuously maintained the first article of his creed, — "I believe in the supremacy of Prussia;" henceforth he upheld with equal vigor the second, — "I believe in the autocracy of the King."

The narrow Constitution limited the King's authority, making it coequal with that of the Upper and Lower Chambers, but Bismarck quickly taught the deputies that he would not allow "a sheet of paper" to intervene between the royal will and its fulfilment. Year after year the Lower House refused to vote the army budget; year after year Bismarck and his master pushed forward the military organization, in spite of the deputies. Noah was not more unmoved by those who came and scoffed at his huge, expensive, apparently useless ark than were the Prussian minister and his King by their critics, who did not see the purpose of the ark the two were building. Bismarck merely insisted that the army, on which depended the integrity of the nation, could not be subjected to the caprice of parties; it was an institution above parties, above politics, he said, which the King alone must control.

At the same time, the Minister-President actively pursued his other project,— the expulsion of Austria from Germany. When the King of Denmark died, in December, 1863, the succession to the duchies of Schleswig and Holstein was disputed. Bismarck seized the occasion for occupying the disputed territory, in partnership with Austria. England protested, France muttered, but neither cared to risk a war with the allied robbers. When it came to dividing the spoils, Bismarck, who had recently gauged Austria's strength, struck for the lion's share. Austria resisted. Bismarck then approved himself a master of diplomacy. Never was he more clever or more unscrupulous, shifting from argument to argument, delaying the open rupture till Prussia was quite ready, feigning willingness to submit the dispute to European arbitration while secretly stipulating conditions which foredoomed arbitration to failure, and invariably giving the impression that Austria refused to be conciliated. As the juggler lets you see the card he wishes you to see, and no other, so Bismarck always kept in full view, amid whatever shuffling of the pack, the apparent legality of Prussia. In the end he drove Austria to desperation.

In June, 1866, war came, with fury. One Prussian army crushed with a single blow the

German states which had promised to support Austria; another marched into Bohemia, and, in seven days, confronted the imperial forces at Sadowa. There was fought a great battle, in which the Prussian crown prince repeated the master stroke of Blücher at Waterloo, and then Austria, hopelessly beaten, sued for peace.

Bismarck now showed himself astute in victory. Having ousted Austria from Germany, he had no wish to wreak a vengeance that she could not forgive. Taking none of her provinces, he exacted only a small indemnity. With the German states he was equally discriminating: those which had been inveterately hostile he annexed to Prussia; the others he let off with a fine. He set up the North German Confederation, embracing all the states north of the river Main, in place of the old German Confederation; and thus Prussia, which had now two thirds of the population of Germany, was undisputed master. The four South German states, Bavaria, Würtemberg, Hesse, and Baden, signed a secret treaty, by which they gave the Prussian King the command of their troops in case of war.

Europe, which had witnessed with astonishment these swift proceedings, understood now that a great reality had arisen, and that Bismarck was its heart. In France, surprise gave way to indignation. Were not the French the arbiters of Europe? How had it happened that their Emperor had permitted a first-rate power to organize without their consent? Napoleon III, who knew that his sham empire could last only so long as he furnished his restless subjects food for their vanity, strove to convince them that he had not been outwitted; that he still could dictate terms. He demanded a share of Rhineland to offset Prussia's aggrandizement; Bismarck refused to cede a single inch. Napoleon bullied; Bismarck published the secret compact with the South Germans. Napoleon forthwith decided that it was not worth while to go to war.

We have all heard of the sportsman who boasted of always catching big strings of fish. But one day, after whipping every pool and getting never a trout, he was fain, on his way home, to stop at the fishmonger's and buy a salt herring for supper. Not otherwise did Napoleon, who had been very forward in announcing that he would *take* land wherever he chose, now stoop to offer to *buy* enough to appease his greedy countrymen. He would pay ninety million francs for Luxemburg, and the King of Holland, to whom it belonged, was willing to sell at that price; but Bismarck would consent only to withdraw the Prussian garrison from the grand duchy, after destroying the fortifications, and to its conversion into a neutral state. That was the sum of the satisfaction Napoleon and his presumptuous Frenchmen got from their first encounter. A few years before, Napoleon, who had had frequent interviews with Bismarck and liked his joviality, set him down as "a not serious man;" whence we infer that the Emperor was a dull reader of character.

Although, by this arrangement, the Luxemburg affair blew over, neither France nor Prussia believed that their quarrel was settled. Deep in the heart of each, instinct whispered that a life-and-death struggle was inevitable. Bismarck, amid vast labor on the internal organization of the kingdom, held Prussia ready for war. He would not be the aggressor, but he would decline no challenge.

In July, 1870, France threw down the glove. When the Spaniards elected Prince Leopold of Hohenzollern to their vacant throne, France demanded that King William should compel Leopold to resign. William replied that, as he had not influenced his kinsman's acceptance, he should not interfere. The prince, who was not a Prussian, withdrew of his own accord. But the French Secretary of State, the Duc de Gramont, had blustered too loudly to let the matter end without achieving his purpose of humbling the Prussian

King. He therefore telegraphed Benedetti, the French Ambassador, to force King William to promise that at no future time should Leopold be a candidate for the Spanish crown. Benedetti delivered his message to William in the public garden at Ems; and William, naturally refusing to bind himself, announced that further negotiations on the subject would be referred to the Foreign Minister.

The following morning Bismarck published a dispatch containing a brief report of the interview; adding, however, that the King "declined to receive the French Ambassador again, and had him told by the adjutant in attendance that his Majesty had nothing further to communicate to the Ambassador." This deceitful addition produced exactly the effect which Bismarck intended: every German, whether Prussian or not, was incensed to learn that the representative German King had been hectored by the French emissary, and every Frenchman was enraged that the Prussian King had insulted the envoy of the "grand nation." Bismarck, who had feared that another favorable moment for war was passing, now exulted, and Moltke, who had for years been carrying the future campaign in his head, and whose face grew sombre when peace seemed probable, now smiled a grim, contented smile. In Paris, the ministers, the deputies, the newspapers, and the populace clamored for war. Apparently, Napoleon alone felt a slight hesitation; but he could hesitate no longer when the popular demand became overwhelming. On July 19 France made a formal declaration of war, and the Parisians laid bets that their victorious troops would celebrate the Fête Napoléon — August 15 — in Berlin. Had not their War Minister, Lebœuf, assured them that everything was ready, down to the last button on the last gaiter of the last soldier?

We cannot describe here the terrible campaign which followed. In numbers, in equipment, in discipline, in generalship, in everything but bravery, the French were quickly outmatched. When Napoleon groped madly for some friendly hand to stay his fall, he found that Bismarck had cut off succor from him. The South Germans, whom the French had hoped to win over, fought loyally under the command of Prussia; Austria, who might have been persuaded to strike back at her late conqueror, dared not move for fear of Russia, whose friendship Bismarck had secured; and Italy, instead of aiding France, lost no time in completing her own unification by entering Rome when the French garrison was withdrawn. Forsaken and outwitted, the French Empire sank without even an expiring flash of that tinsel glory which had so long bedizened its corruption. And when the French people, lashed to desperation, continued the war which the Empire had brought upon them, they but suffered a long agony of losses before accepting the inevitable defeat. They paid the penalty of their former arrogance in every coin known to the vanquished, — in military ruin, in an enormous indemnity, in the occupation of their land by the victorious Prussians, and in the cession of two rich provinces. Nor was that enough: they had to submit to a humiliation which, to the imagination at least, seems the worst of all, — the proclamation of the Prussian King William as German Emperor in their palace at Versailles, the shrine of French pomp, where two centuries before Louis XIV had embodied the ambition, the glory, and the pride of France. The German cannon bombarding beleaguered Paris paused, while the sovereigns of the German states hailed William as their Emperor.

This consummation of German unity was the logical outcome of an international war, in which all the Germans had been impelled, by mutual interests quite as much as by kinship, to join forces against an alien foe. Twenty years before, Bismarck had opposed German unity, because it would then have made Prussia the plaything of her confederates; in this later scheme he was the chief agent, if not the originator, for he knew that the primacy of Prussia ran no more risk.

Let us pause a moment and look back. Only a decade earlier, in 1861, when Bismarck became minister, Prussia was but a second-rate power, Germany was a medley of miscellaneous states, Austria still held her traditional supremacy, the French Emperor seemed firmly established. Now, in 1871, Austria has been humbled, France crushed, Napoleon whiffed off into outer darkness, and Prussia stands unchallenged at the head of United Germany. Many men — the narrow, patient King, the taciturn Moltke, the energetic Von Roon — have contributed to this result; but to Bismarck rightly belongs the highest credit. Slow to prepare and swift to strike, he it was who measured the full capacity of that great machine, the Prussian army, and let it do its work the moment Fortune signaled; he it was who knew that needle guns and discipline would overcome in the end the long prestige of Austria and the wordy insolence of France. Looking back, we are amazed at his achievements, — many a step seems audacious; but if we investigate, we find that Bismarck had never threatened, never dared, more than his strength at the time warranted. The gods love men of the positive degree, and reward them by converting their words into facts.

Of the German Empire thus formed Bismarck was Chancellor for twenty years. His foreign policy hinged on one necessity, — the isolation of France. To that end he made a Triple Alliance, in which Russia and Austria were his partners first, and afterward Italy took Russia's place. He prevented the Franco-Russian coalition, which would place Germany between the hammer and the anvil. From 1871 to 1890 he was not less the arbiter of Europe than the autocrat of Germany.

Nevertheless, although in the management of home affairs Bismarck usually prevailed, he prevailed to the detriment of Germany's progress in self-government. The Empire, like Prussia herself, is based on constitutionalism: what hope is there for constitutionalism, when at any moment the vote of a majority of the people's representatives can be nullified by an arbitrary prime minister? Bismarck carried his measures in one of two ways: he either formed a temporary combination with mutually discordant parliamentary groups and thereby secured a majority vote, or, when unable to do this, by threatening to resign he gave the Emperor an excuse for vetoing an objectionable bill. Despising representative government, with its interminable chatter, its red tape, its indiscreet meddling, and its whimsical revulsions,

Bismarck never concealed his scorn. If he believed a measure to be needed, he went down into the parliamentary market-place, and by inducements, not of money, but of concessions, he won over votes. At one time or another, every group has voted against him and every group has voted for him. When he was fighting the Vatican, for instance, he conciliated the Jews; when Jew-baiting was his purpose, he promised the Catholics favor in return for their support. Being amenable to the Emperor alone, and not, like the British premier, the head of a party, he dwelt above the caprice of parties. Men thought, at first, to stagger him by charges of inconsistency, and quoted his past utterances against his present policy. He laughed at them. Consistency, he held, is the clog of men who do not advance; for himself, he had no hesitation in altering his policy as fast as circumstances required. With characteristic bluntness, he did not disguise his intentions. "I need your support," he would say to a hostile group, "and I will stand by your bill if you will vote for mine." "*Do ut des*" was his motto; "an honest broker" his self-given nickname.

Such a government cannot properly be called representative; it dangles between the two incompatibles, constitutionalism and autocracy. Doubtless Bismarck knew better than the herd of deputies what would best serve at a given moment the interests of Germany; but his methods were demoralizing, and so personal that they made no provision for the future. His system could not be permanent unless in every generation an autocrat as powerful and disinterested as himself should arise to wield it; but nature does not repeat her Bismarcks and her Cromwells. At the end of his career, Germany has still to undergo her apprenticeship in self-government.

Two important struggles, in which he engaged with all his might, call for especial mention.

The first is the *Culturkampf*, or contest with the Pope over the appointment of Catholic bishops and clergy in Prussia. Bismarck insisted that the Pope should submit his nominations to the approval of the King; Pius IX maintained that in spiritual matters he could be bound by no temporal lord. Bismarck passed stern laws; he withheld the stipend paid to the Catholic clergy; he imprisoned some of them; he broke up the parishes of others. It was the mediaeval war of investitures over again, and again the Pope won. Bismarck discovered that against the intangible resistance of Rome his Krupp guns were powerless. After fifteen years of ineffectual battling, the Chancellor surrendered.

Similar discomfiture came to him from the Socialists. When he entered upon his ministerial career, they were but a gang of noisy fanatics; when he quitted it, they were a great political party, holding the balance of power in the Reichstag, and infecting Germany with their doctrines. At first he thought to extirpate them by violence, but they throve under persecution; then he propitiated them, and even strove to forestall them by adopting Socialistic measures in advance of their demands. If the next epoch is to witness the triumph of Socialism, as some predict, then Bismarck will surely merit a place in the Socialists' Saints' Calendar; but if, as some of us hope, society revolts from Socialism before experience teaches how much insanity underlies this seductive theory, then Bismarck will scarcely be praised for coquetting with it. For Socialism is but despotism turned upside down; it would substitute the tyranny of an abstraction — the state — for the tyranny of a personal autocrat. It rests on the fallacy that though in every individual citizen there is more or less imperfection, — one dishonest, another untruthful, another unjust, another greedy, another licentious, another willing to grasp money or power at the expense of his neighbor, — yet by adding up all these units, so imperfect, so selfish, and calling the sum "the state," you get a perfect and unselfish organism, which will manage without flaw or favor the whole business, public, private, and mixed, of mankind. By what miracle a coil of links, separately weak, can be converted into an unbreakable chain is a secret which the prophets of this Utopia have never condescended to reveal. Not more state interference, but less, is the warning of history.

The fact which is significant for us here is that Socialism has best thriven in Germany, where, through the innate tendency of the Germans to a rigid system, the machinery of despotism has been most carefully elaborated, and where the interference of the state in the most trivial affairs of life has bred in the masses the notion that the state can do everything, — even make the poor rich, if they can only control the lever of the huge machine.

Nevertheless, though Bismarck has been worsted in his contest with religious and social ideas, his great achievement remains. He has placed Germany at the head of Europe, and Prussia at the head of Germany. Will the German Empire created by him last? Who can say? The historian has no business with prophecy, but he may point out the existence in the German Empire to-day of conditions that have hitherto menaced the safety of nations. The common danger seems the strongest bond of union among the German states. De feat by Russia on the east or by France on the west would mean disaster for the South Germans not less than for the Prussians; and this peril is formidable enough to cause the Bavarians, for instance, to fight side by side with the Prussians. But there can be no homogeneous internal government, no compact nation, so long as twenty or more dynasties, coequal in dignity though not in power, flourish simultaneously. Historically speaking, Germany has never passed through that stage of development in which one dynasty swallows up its rivals, — the experience of England, France, and Spain, and even of polyglot Austria.

Again, Germany embraces three unwilling members, — Alsace-Lorraine, Schleswig, and Prussian Poland, — any of which may serve as a provocation for war, and must remain a constant source of racial antipathy. How grievous such political thorns may be, though small in bulk compared to the body they worry, England has learned from Ireland.

Finally, if popular government — the ideal of our century — is to prevail in Germany, the despotism extended and solidified by Bismarck will be swept away. Possibly, Germany could not have been united, could not have humbled Austria and crushed France, under a Liberal system; but will the Germans forever submit to the direction of an iron chancellor, or glow with exultation at the truculence of a strutting autocrat who flourishes his sword and proclaims, "My will is law"? No other modern despotism has been so patriotic, honest, and successful as that of Bismarck; but will the Germans never awake to the truth that even the best despotism convicts those who bow to it of a certain ignoble servility? Or will they, as we have suggested, transform the tyranny of an autocrat into the tyranny of Socialism? We will not predict, but we can plainly see that Germany, whether in her national or in her constitutional condition, has reached no stable plane of development.

And now what shall we conclude as to Bismarck himself? The magnitude of his work no man can dispute. For centuries Europe awaited the unification of Germany, as a necessary step in the organic growth of both. Feudalism was the principle which bound Christendom together during the Middle Age; afterward, the dynastic principle operated to blend peoples into nations; finally, in our time, the principle of nationality has accomplished what neither feudalism nor dynasties could accomplish, the attainment of German unity. In type, Bismarck belongs with the Charlemagnes, the Cromwells, the Napoleons; but, unlike them, he wrought to found no kingdom for himself; from first to last he was content to be the servant of the monarch whom he ruled. As a statesman, he possessed in equal mixture the qualities of lion and of fox, which Machiavelli long ago declared indispensable to a prince. He had no scruples. What benefited Prussia and his King was to him moral, lawful, desirable; to them he was inflexibly loyal; for them he would suffer popular odium or incur personal danger. But whoever opposed them was to him an enemy, to be overcome by persuasion, craft, or force. We discern in his conduct toward enemies no more regard for morality than in that of a Mohawk sachem toward his Huron foe. He might spare them, but from motives of policy; he might persecute them, not to gratify a thirst for cruelty, but because he deemed persecution the proper instrument in that case. His justification would be that it was right that Prussia and Germany should hold the first rank in Europe. The world, as he saw it, was a field in which nations maintained a pitiless struggle for existence, and the strongest survived; to make his nation the strongest was, he conceived, his highest duty. An army of puny-bodied saints might be beautiful to a pious imagination, but they would fare ill in an actual conflict with Pomeranian grenadiers.

Dynamic, therefore, and not *moral*, were Bismarck's ideals and methods. To make every citizen a soldier, and to make every soldier a most effective fighting machine by the scientific application of diet, drill, discipline, and leadership, was Prussia's achievement, whereby she prepared for Bismarck an irresistible weapon. In this application of science to control with greater exactness than ever before the movements of large masses of men in war, and to regulate their actions in peace, consists Prussia's contribution to government; in knowing how to use the engine thus constructed lies Bismarck's fame. When Germans were building air-castles, and, conscious of their irresolution, were asking themselves, "Is Germany Hamlet?" Bismarck saw both a definite goal and the road that led to it. The sentimentalism which has characterized so much of the action of our time never diluted his tremendous will. He held that by blood and iron empires are welded, and that this stern means causes in the end less suffering than the indecisive compromises of the sentimentalists. Better, he would say, for ninety-nine men to be directed by the hundredth man who knows than for them to be left a prey to their own chaotic, ignorant, and internecine passions. Thus he is the latest representative of a type which flourished in the age when the modern ideal of popular government had not yet risen.

How much of his power was due to his unerring perception of the defects in popular government as it has thus far been exploited, we have already remarked.

The Germans have not yet perceived that one, perhaps the chief source of his success was his un-German characteristics. He would have all Germany bound by rigid laws, but he would not be bound by them himself. He encouraged his countrymen's passion for conventionality and tradition, but remained the most unconventional of men. Whatever might complete the conversion of Germany into a vast machine he fostered by every art; but he, the engineer who held the throttle, was no machine. In a land where everything was done by prescription, the spectacle of one man doing whatever his will prompted produced an effect not easily computed. Such characteristics are un-German, we repeat, and Bismarck displayed them at all times and in all places. His smoking a cigar in the Frankfort Diet; his opposition to democracy, when democracy was the fashion; his resistance to the Prussian Landtag; his arbitrary methods in the German Parliament, — these are but instances, great or small, of his un-German nature. And his relations for thirty years with the King and Emperor whom he seemed to serve show a similar masterfulness. A single anecdote, told by himself, gives the key to that service.

At the battle of Sadowa King William persisted in exposing himself at short range to the enemy's fire. Bismarck urged him back, but William was obstinate. "If not for yourself, at least for the sake of your minister, whom the nation will hold responsible, retire," pleaded Bismarck. "Well, then, Bismarck, let us ride on a little," the King at last replied. But he rode very slowly. Edging his horse alongside of the King's mare, Bismarck gave her a stout kick in the haunch. She bounded forward, and the King looked round in astonishment. "I think he saw what I had done," Bismarck added, in telling the story, "but he said nothing."

On Bismarck's private character I find no imputed stain. He did not enrich himself by his office, that hideous vice of our time. He did not, like both Napoleons, convert his palace into a harem; neither did he tolerate nepotism, nor the putting of incompetent parasites into responsible positions as a reward for party service. That he remorselessly crushed his rivals let his obliteration of Count von Arnim witness. That he subsidized a "reptile press," or employed spies, or hounded his assailants, came from his belief that a statesman too squeamish to fight fire with fire would deserve to be burnt. Many orators have excelled him in grace, few in effectiveness. Regarding public speaking as one of the chief perils of the modern state, because it enables demagogues to dupe the easily swayed masses, he despised rhetorical artifice. His own speech was un-German in its directness, un-German in its humor, and it clove to the heart of a question with the might of a battle-axe, — as, indeed, he would have used a battle-axe itself to persuade his opponents, five hundred years ago. Since Napoleon, no other European statesman has coined so many political proverbs and apt phrases. His letters to his family are delightfully natural, and reveal a man of keen observation, capable of enjoying the wholesome pleasures of life, and brimful of common sense, which a rich gift of humor keeps from the dulness of Philistines and the pedantry of doctrinaires. His intercourse with friends seems to have been in a high degree jovial.

Not least interesting to a biographer are those last years of Bismarck's life, between March, 1890, and his death, on July 30, 1898, which he passed in eclipse. To be dismissed by a young sovereign who, but for him, might have been merely a petty German prince, — to be told that he, the master throne-maker, was unnecessary to the callow apprentice, — galled the Titan's heart.

Eight years he was destined to endure this mortification; and although his countrymen everywhere hailed him as their hero, the fact of dismissal gave him no repose. Europe has seen no similar spectacle since she bound Napoleon, Prometheus-like, on St. Helena. But Napoleon, chafing his life away there, had at least the satisfaction of reflecting that it took all Europe, allied with Russia's blizzards and Spain's heats, to conquer him. Bismarck, storming in his exile from power, felt now scorn, now hate, for the "young fellow" (as he called him) who had turned him out. Here, if ever, Nemesis showed her work. Bismarck's whole energy had been bent for fifty years on fortifying the autocracy of the Prussian monarchs; and now a young autocrat run from this mould bade him go — and he went. We may believe that it did not solace Bismarck to find that the "young fellow" could get on without him; or to see that in England Gladstone, six years his elder, led his nation till long past eighty; Gladstone, — whom he had so often jeered at as an empty rhetorician, — England, which he despised as the home of representative government. Could it be that constitutionalism was kinder than despotism to master statesmen?

A great man we may surely pronounce him, long to be the wonder of a world in which greatness of any kind is rare. If you ask, "How does he stand beside Washington and Lincoln?" it must be admitted that his methods would have made them blush, but that his patriotism was not less enduring than theirs. With the materials at hand he fashioned an empire; it is futile to speculate whether another, by using different tools, could have achieved the same result. Bismarck knew that though his countrymen might talk eloquently about liberty, they loved to be governed; he knew that their genius was mechanical, and he triumphed by directing them along the line of their genius. He would have failed had he appealed to the love of liberty, by appealing to which Cavour freed Italy; or to the love of glory, by appealing to which Napoleon was able to convert half of Europe into a French province. Bismarck knew that his Prussians must be roused in a different way.

It may be that the empire he created will not last; it is certain that it cannot escape modifications which will change the aspect he stamped upon it; but we may be sure that, whatever happens, the recollection of his Titanic personality will remain. He belongs among the giants, among the few in whom has been stored for a lifetime a stupendous energy, — kinsmen of the whirlwind and the volcano, — whose purpose seems to be to amaze us that the limits of the human include such as they. At the thought of him, there rises the vision of mythic Thor with his hammer, and of Odin with his spear; the legend of Zeus, who at pleasure held or hurled the thunderbolt, becomes credible.

NAPOLEON III

Madame de Staël said of Rienzi and his Romans: "They mistook reminiscences for hopes;" of the second French Empire and the third Napoleon we may say: "They staked their hopes on reminiscences."

In our individual lives we realize the power of memory, suggestion, association. If we have ever yielded to a vice, we have felt, it may be years after, how the sight of the old conditions revives the old temptation. A glance, a sound, a smell, may be enough to conjure up a long series of events, whether to grieve or to tempt us, with more than their original intensity. So we learn that the safest way to escape the enticement is to avoid the conditions. Recent psychology has at last begun to measure the subtle power of suggestion.

But now, suppose that instead of an individual a whole nation has had a terrific experience of succumbing to temptation, and that a cunning, unscrupulous man, aware of the force of association and reminiscence, deliberately applies both to reproduce those conditions in which the nation first abandoned itself to excess: the case we have supposed is that of France and Louis Napoleon. Before the reality of their story the romances of hypnotism pale.

After Sédan it was the fashion to regard Louis Napoleon as the only culprit in the gilded shame of the Second Empire; we shall see, however, that the great majority of Frenchmen longed for his coming, applauded his victories, and by frequent vote sanctioned his deeds. A free people keeps no worse ruler than it deserves.

The Napoleonic legend, by which Louis Napoleon rose to power, was not his creation, but that of the French: he was simply shrewd, and used it. What was this legend?

When allied Europe finally crushed the great Napoleon at Waterloo, France breathed a sigh of relief. Twenty campaigns had left her exhausted: she asked only for repose. This the Restoration gave her. But the gratification of our transient cravings, however strong they may be, cannot long satisfy; and when the French recovered from their exhaustion, they felt their permanent cravings return. The Bourbons, they soon realized, could not appease those dominant Gallic desires. For the Bourbons had destroyed even that semblance of liberty Napoleon took care to preserve; they persecuted democratic ideas; they brought back the old aristocracy, with its mildewed haughtiness; they babbled of divine right, — as if the worship of St. Guillotine had not supervened. During twenty years France had been the arbitress of Europe; now, under the narrow, forceless Bourbons, she was treated like a second-rate power. Waterloo had meant not only the destruction of Napoleon, from which France derived peace, but also humiliation, which galled Frenchmen more and more as their normal sensitiveness returned.

The Bourbons, knowing that they might be tolerated so long as they were not despised, got up a military promenade into Spain, to prove that France could still meddle in her neighbors' affairs, and that the Bourbons were not less mighty men of war than the Bonapartes. They captured the Trocadero, and restored vile King Ferdinand and his twenty-six cooks to the throne of Spain; and they hoped that the one-candle power of fame lighted by these exploits would outdazzle the Sun of Austerlitz. But no, the dynasty of Bourbon, long since headless, proved to be rootless too: one evening Charles X played his usual game of whist at St. Cloud; the next, he was posting out of France with all the speed and secrecy he could command.

Louis Philippe, who came next, might have been

expected to please everybody: Royalists, because he was himself royal; Republicans, because he was Philippe Egalité's son; constitutionalists, because he hated autocratic methods: shopkeepers of all kinds, because he was 'practical.' And in truth his administration may be called the Golden Age of the *bourgeoisie*,—the great middle class which, in France and elsewhere, was superseding the old aristocracy. Napoleon had organized a nobility of the sword; after him came the nobility of the purse. Louis Philippe could say that under his rule France prospered: her merchants grew rich; her factories, her railroads, all the organs of commerce, were healthily active. And yet she was discontented. The spectacle of her Citizen King walking unattended in the streets of Paris, his plump thighs encased in democratic trousers, his plump and ruddy face wearing a complacent smile, his whole air that of the senior partner in some old, respectable, and rich firm,— even this failed to satisfy Frenchmen. "He inspires no more enthusiasm than a fat grocer," was said of him. Frenchmen did not despise money-making, but they wanted something more: they wanted *gloire*.

Let us use the French word, because the English *glory* has another meaning. *Glory* implies something essentially noble,—nay, in the Lord's

Prayer it is a quality attributed to God himself: but *gloire* suggests vanity; it is the food braggarts famish after. The minute-men at Concord earned true glory; but when the United States, listening to the seductions of evil politicians, attacked and blasted a decrepit power, — fivefold smaller in population, twenty-fold weaker in resources, — they might find *gloire* among their booty, but glory, never. As prosperity increased, the Gallic appetite for *gloire* increased. Louis Philippe made several attempts to allay it, but he dared not risk a foreign war, and the failure of his attempts made him less and less respected.

And now arose the Napoleonic legend, at first no more than a bright exhalation in the evening, but gradually taking on the sweep, the definiteness, the fascination, of mirage at noonday. Time enough had elapsed to dull or quite blot out the recollections of the hardships and strains, the millions of soldiers killed and wounded, the taxes, the grievous tyranny; men remembered only the victories, the rewards, and the splendor. A new generation, unacquainted with the havoc of war, had grown up, to listen with fervid envy to the reminiscences of some gray-haired veteran, who had made the great charge at Wagram or ridden behind Ney at Borodino. Those exploits were so stupendous as to seem incredible, and yet they were vouched for by too many survivors to be doubted. Was not Thiers setting forth the marvelous story in nineteen volumes? Were not Béranger and even Victor Hugo singing of the departed grandeur? Were not the booksellers' shelves loaded with memoirs, lives, historical statements, polemics? Paris, France, seemed to exist merely to be the monument of one man. And wherever the young Frenchman traveled — in Spain, in Italy, along the Rhine or the Danube, to Vienna, or Cairo, or Moscow — he saw the footprints of French valor and French audacity, reminders that Napoleon had made France the mistress of Europe. No Frenchman, were he Bourbon or Republican, but felt proud to think that his countrymen had humbled Prussia and Austria.

Confronted by such recollections, the France of Louis Philippe looked degenerate. It offered nothing to thrill at, to brag over; it sinned in having — what it could not help — a stupendous past just behind it. So the Napoleonic legend grew. The body of the great Emperor was brought home from St. Helena, to perform more miracles than the mummy of a mediæval saint. Power and *gloire* came to be regarded as the products of a Napoleonic régime: to secure them it was only necessary to put a Bonapartist on the throne.

Contemporaneous with the expansion of this spell, Socialism grew up, and taught that, just as the *bourgeoisie* had overthrown the old privileged classes in the French Revolution, so now the working classes must emancipate themselves from the tyranny of the *bourgeoisie*. Political equality without industrial equality seemed a mockery. In this wise the doctrines of a score of Utopians penetrated society to loosen old bonds and embitter class with class. And besides all this, there was the usual wrangle of political parties. The tide of opposition rose, and on February 24, 1848, swept away Louis Philippe and his minister Guizot. Among the many fortune-seekers whom that tide brought to land was Louis Napoleon.

He was born in Paris, April 20, 1808, his mother being Hortense Beauharnais, who had married Louis Bonaparte, King of Holland. The younger Louis could just remember being petted in the Tuileries by the great Emperor: then, like all the Bonapartes, he had been packed off into exile. His youth was chiefly spent on Lake Constance, at Augsburg, and at Thun. In 1831 he had joined the Carbonari plotters in Italy. The next year, through the death of his elder brother and of the great Napoleon's son, he became the official Pretender to the Bonapartist hopes. People knew him only as a visionary, who talked much about his "star," and by writings and deeds tried to persuade the world that he too, like his uncle, was a man of destiny. A few adventurers gathered round him, eager to take the one chance in a thousand of his success. Accompanied by some of these, in 1836, he appeared before the French troops at Strasburg, expecting to be acclaimed Emperor and to march triumphantly to Paris. He did go to Paris, escorted by policemen; but his attempt seemed so foolish that Louis Philippe merely paid his passage to America to be rid of him.

The Prince soon returned to Europe and settled in London, where he lived the fast life of the average nobleman. In 1840 he set out on another expedition against France. Carrying a tame eagle with him, he landed at Boulogne: but again neither the soldiers nor populace welcomed him; the eagle seems to have been a spiritless fowl, likewise incapable of arousing enthusiasm; and the Prince shortly after was under imprisonment for life in the fortress of Ham. Nearly six years later he bribed a jailer, escaped to London, and, like Micawber, waited for something to turn up.

The fall of Louis Philippe gave Prince Louis his opportunity. He hurried to Paris, but was considerate, or cunning, enough to hold aloof for a while from disturbing public affairs. In those first months of turmoil many aspirants were destroyed, by their own folly and by mutual collision. Discreetly, therefore, he stood aside and watched them disappear.

Of the several factions, the Socialists and Red Republicans first profited by the Revolution. They organized that colossal folly, the National Workshops, in which 120,000 loafers received from the state good wages for pretending to do work which, had they done it, would have benefited no one. When the state, realizing that it could not continue this preposterous expense, proposed to close the workshops, the loafers became sullen: when the wages were cut off, they throttled Paris. For four days, in June, 1848, they made the streets of Paris their battle-ground, and succumbed only after 30,000 of their number had been killed, wounded, or captured by Cavaignac's troops. The terror inspired by those idlers of Louis Blanc's workshops was the corner-stone of the Second Empire.

A few weeks later, Louis Napoleon, elected by five constituencies, took his seat in the Assembly. His uncle's name was still his only political capital. His own record — the Strasburg and Boulogne episodes — inspired mirth. In person there was nothing commanding about him. An "olive-swarthy paroquet" some one called him. "His gray eyes," says De Tocqueville, " were dull and opaque, like those thick bull's-eyes which light the stateroom of a ship, letting the light pass through, but out of which we can see nothing." In after years " inscrutable" was the word commonly chosen to describe his cold, unblinking gaze. Reserve always characterized his manners; for even when most affable, his intimates felt that he concealed something or simulated something.

In the Assembly he strove for no sudden recognition; outside, however, he and his emissaries busied themselves night and day fanning the embers of Imperialism; and when, in December, 1848, the French people voted for a president, Louis Napoleon received 5,434,000 votes, while Cavaignac, his nearest competitor, had but 1,448,000. How had this come about? Old soldiers and peasants composed the great bulk of his supporters, every one of them glad to vote for "the nephew of the Emperor." Next, Socialists, blue blouses and others, voted for him because they hated Cavaignac for repressing Red Republicanism in June; and Monarchists of both stripes, believing that he would be an easy tool for their plots, preferred him to the unyielding Cavaignac. Mediocrity and other negative qualities thus availed to transform Louis the Ridiculed into the first President of the Republic. " We made two blunders in the case of Louis Napoleon," said Thiers; "first in deeming him a fool, and next in deeming him a genius." Louis Napoleon knew not only how to profit by both of these blunders, but also how to superinduce either belief in the French mind.

Having sworn to uphold the Republic, he began his administration. During several months he let no sign of his ambition flutter into view, but seemed wholly bent on discharging the duties of president. In the spring of 1849, however, he put forth a feeler by engineering the expedition against the Roman Republic. Honest Frenchmen protested, but a majority in the Assembly supported him; and presently the instinct to be revenged on the Romans for defending themselves, and thereby inflicting losses on the French, silenced many who had disapproved of the expedition at the outset. Only the Radicals forcibly resisted, but their revolt was quickly put down. Louis Napoleon gained the prestige of having successfully reasserted French influence in Italy, where, for a generation, it had been supplanted by the influence of Austria. Furthermore, by becoming guardian of the Pope, he propitiated the Clericals, who might some time be useful. That he also roused the wrath of the Red Republicans did not spoil his prospects.

One year, two years passed. Faction discredited faction. Every one looked on the Republic as but a preparation for either Anarchy or the Empire. The Reds, irreconcilable and ferocious, terrorized the imagination of every one else. No doubt the majority of honest Frenchmen — if by honest we mean the really intelligent and patriotic minority — wished a republic, but those Red Extremists had made all Republicans indiscriminately odious; and as the Royalist plotters showed neither courage nor ability, the great multitude of Frenchmen came to regard the Empire or Anarchy as their only alternatives. Most of them, having nothing to gain through disorder, leaned to the side which promised to leash the bloodhounds of murder and pillage. Spasm after spasm of terror swept over Paris, and when Paris shudders in the evening the rest of France shudders by daybreak. Anything to prevent the triumph of the Reds — with their guillotine and their abolition of private ownership of property — became the ruling instinct of all other Frenchmen.

Louis Napoleon, we may be sure, took care to encourage the belief that he alone could save France from the abyss. In addition to his recognized newspaper organs, he employed a literary bureau to spread broadcast his portrait, his biography, and even songs with an Imperialist refrain. He knew the political persuasiveness of cigars and sausages distributed among the troops, and of wine dispensed to their officers. He was by turns modest — declaring that his sole purpose was to obey the Constitution — and bold, announcing that he would not shrink from making France strong and prosperous, whenever Frenchmen intrusted that task to him. In his capacity for waiting, he gave the best proof of his ability; and we must add that the Assembly, by its folly, gave him indispensable aid.

The Assembly, for instance, restricted the suffrage, in the hope that, by preventing workmen from voting, the victory of the Reds might be staved off. Again, the Constitution declared that no president was eligible for reëlection until he had been four years out of office. As the time for thinking of Louis Napoleon's successor approached, the moderates of all parties urged that the Constitution be amended, so that he might be quietly reëlected, — there being no other candidate who promised to preserve order. But the factious deputies, by a narrow vote, rejected the amendment.

Napoleon now saw his chance, and openly assailed the Assembly. He posed as the champion of universal suffrage, the true representative of the people misrepresented by the factious deputies. They proposed to subject France to the uncertainties of a political campaign: his continuation in office would mean the certain maintenance of order. But Napoleon did not rely on demagogy alone: in secret he plotted a *coup d'état*.

The trade of house or bank burglar long ago fell into disrepute: not so that of the state burglar, who, if he succeed, may wear ermine jauntily, — for ermine, like charity, covers a multitude of sins. Louis Napoleon, ready to risk everything, laid his plans for stealing the government of France. The venture was less difficult than it seems, for if he could win over four or five men the odds would be with him. He must have the Prefect of Paris, the Commandant of the Garrison, the Ministers of War and of the Interior: others might make assurance double sure, but these were absolutely necessary.

Early in the spring of 1851 he set to work. Chief among his accomplices was his half brother, Morny, — a facile, audacious man, whose reputation, if he had ever had any, would have been lost long since in stock-swindling schemes; after him, in importance, came Persigny, an adventurer who had fastened on Louis Napoleon fifteen years before; Fleury, a major most active and efficient, without qualm, for he foresaw a marshal's *bâton ;* and Maupas, one of those easy villains who, never having been suspected of honesty, are spared the fatigue of pretending to be better than they are. If we assume that all these gentlemen were Imperialists for revenue only, we shall do them no injustice.

Their first move was to send Fleury to Algiers to secure a general to act as minister of war. He had not to search long; for Saint Arnaud, one of the Algerian officers, guessing Fleury's purpose, offered his services forthwith. But Saint Arnaud stood only fifty-third in the line of promotion among French generals; some excuse must be found for passing by his fifty-two seniors. In a few weeks the French press and official gazette announced an outbreak of great violence among the Kabyles in Algeria; a little later they reported that the insurrection had been subdued by the energy of General Saint Arnaud; then, another proper interval elapsing, Saint Arnaud had come to Paris as minister of war.

It took less trouble to dismiss the Prefect of the Seine, and to substitute Maupas for him. Magnan, who commanded the troops, had already been corrupted. Half-brother Morny, at the critical moment, would appear in the Ministry of the Interior. The National Guard and the Public Printer could both be counted on, — the latter required for the prompt issuing of manifestoes.

Everything being ready, the President, after some brief delays, set December 2 — the anniversary of Austerlitz, and of the coronation of the great Napoleon — for committing the crime.

On the evening of December 1, he held his weekly reception at the Elysée; moved with his habitual courtesy among the guests; seemed less stiff than usual, — as if relieved of a burden; then went to his study for a last conference with his fellow-conspirators. The next morning Paris learned that two hundred leading citizens, military and political, including many deputies, had been arrested and taken to Vincennes. Placards declared that the President, having had news of a plot against the state, had stolen a march on the plotters, dissolved the Assembly, proclaimed universal suffrage, and called for a plebiscite to accept or reject the constitution he would frame. At first, the stupefied Parisians knew not what to do. Then the deputies who had escaped arrest met and voted to depose the President; but his gendarmes quickly broke up the meeting, and lodged the deputies in prison. Thanks to the system of centralization which France had long boasted of, Morny, from the Ministry of the Interior, controlled every prefect in France by telegraph. The provinces were informed that Paris had accepted the *coup d'état* almost before Paris had collected her dazed senses on the morning of the 2d of December.

The chief politicians and other leaders being caged, there was no one left, except among the workingmen, to direct a resistance. They did revolt, and Napoleon and Saint Arnaud gave them free play to raise barricades, to arm and gather. Then the eighty thousand soldiers in Paris surrounded them, stormed their barricades, and made no prisoners. Acompanying this suppression of the mob was the bloodthirsty massacre of a multitude of defenseless men, women, and children who had collected on the boulevards to see the troops move against the barricaders. They were shot down in cold blood, the soldiers (according to general report) having been rendered ferocious by drink. Thus was achieved the crime of the *coup d'état*.

By this crime Napoleon had demonstrated that he had the necessary force to put down the lawless, and that he did not hesitate to use it; by massacring the innocent throng, he made the army his accomplices, against any risk of their fraternizing with the populace. A fortnight later, 7,439,000 Frenchmen ratified his crime and elected him president for ten years: only 646,000 voted against him. Napoleon the Great, by the *coup d'état* of the 18th Brumaire, had suppressed the Directory; his imitative nephew could now point to an equally successful 2d of December.

France acquiesced all the more readily because she was put under martial law. One hundred thousand suspects were arrested, and more than ten thousand were deported to Cayenne and Algeria. Police inquisitions, military commissions, and the other usual devices of tyranny quickly smothered resistance. Under the pretense of suppressing anarchy, — an anarchist meaning any one who did not submit to Louis Napoleon, — persecution supplanted law and justice. Had you asked to see most of the Frenchmen whose names were the most widely known, you would have been told that they were in exile.

Like his uncle, Louis Napoleon waited a little before putting on the purple. Only on December 2, 1852, the anniversary of his crime, did he have himself proclaimed emperor. The mockery of a plebiscite had preceded, and he had assured France and Europe that the "Empire means peace."

Having reached the throne, he made the following arrangements for staying on it. He organized a Senate and a Council of State, whose members he appointed. The public were allowed to elect members to the Corps Législatif, or Legislature; but as his minions controlled the polls, only such candidates as he preferred were likely to be chosen. He suffered a few opponents to be elected, in order to have it appear that he encouraged discussion. Otherwise, he scarcely took pains to varnish his autocracy. As a deft Chinese carver incloses a tiny figure in a nest of ivory boxes, so did Bonaparte imprison the simulacrum of Liberty in the innermost compartment of the political cage in which he held France captive.

What must the condition of the French people have been that they submitted! How much antecedent incapacity for government, how much cherishing of unworthy ideals, were implied by the success of such an adventurer! And what could patriotism mean, when the French fatherland meant the land of Louis Napoleon, Morny, Maupas, Persigny, and their unspeakable underlings? The new Empire gave France what is called a strong government, by which commercially she throve. Tradesmen, seeing business improve and their hoards grow, chafed less at the loss of political freedom. The working classes were propitiated by public works — the favorite nostrum of socialists and tyrants — organized on a vast scale. Pensions were showered on old soldiers, or their widows. Taxes ran high; the public debt had constantly to be increased: but an air of opulence pervaded France.

Established at home, Napoleon now looked abroad for *gloire*. Before his elevation, some one had warned him that he would find the French a very hard people to govern. "Not at all," he replied; "all that they need is a war every four years." Europe had formally recognized him, — no country being more ready than England to condone his great crime. Queen Victoria, the typical British matron, exchanging visits with the Imperial adventurer made an edifying spectacle! Presently the land-greed of England and the *gloire*-thirst of France brought the sons of the Britons who had whipped the great Napoleon at Waterloo into an alliance with the sons of the Frenchmen who had there been whipped; and in the summer of 1854 British and French fleets swept through the Bosphorus and across the Black Sea, and landed two armies near Sebastopol. Of the Crimean war which ensued, we need say no more than that it was immoral in conception, blundering in execution, and ineffectual in results. Nevertheless, it supplied Napoleon III with just what he had sought. He extracted from it large quantities of *gloire*. Marshal's *bâtons* and military promotions, the parade of returning troops, the assembling at Paris of the European envoys who were to agree on a treaty of peace, — what did all this show but that Europe had accepted

Napoleon III at his own valuation? In Russia's wilderness of snow the great Napoleon had been ruined; now his nephew posed as the humbler of Russia. The great Napoleon had been finally crushed by England: now his nephew had enticed good, pious England into an alliance, and thereby he had surely avenged his uncle. The last European compact, humiliating to France, had been signed at Vienna: the new compact, signed at Paris, bore witness to the supremacy of France.

That year 1856 marks the acme of Napoleon the Third's career. It saw him the recognized arbiter of Europe. The world, which worships success, forgot that the suave, impassive master of the Tuileries had been Louis the Ridiculed, a political vagabond and hapless pretender, only ten years before. Now, as arbiter, he would meddle when he chose, and the world should not gainsay him. Moreover, he believed his power so secure that he was willing to forgive those whom he had injured. He had gained what he wanted: why, therefore, should *they* reject his amnesty?

Unscrupulously selfish till he had attained his ends, Napoleon III had, nevertheless, curious streaks of disinterestedness in his nature. What but Quixotism impelled him to promise to free Italy from her bondage to Austria? He might add thereby to his personal renown, but the French people, who must pay the bills and furnish the soldiers, were offered no adequate compensation. Whatever his motives, he crossed the Alps in the spring of 1859, joined the Piedmontese, and defeated the Austrians in two great battles. But after Solferino he paused, grew anxious, and drew back. Many reasons were hinted at: he had been horrified at the sight of twelve thousand corpses festering in the midsummer heat on the battlefield; he perceived that the campaign must last many months before the Austrians could be dislodged from the Quadrilateral; he dreaded to create in Italy a kingdom strong enough to be a menace to France; he was worried at the mobilization of the Prussian army, foreboding a war on the Rhine. Motives are usually composite: perhaps, therefore, all these, and others, made him resolve to quit Italy with his mission only half achieved. But of all his schemes, that Italian expedition has alone escaped the condemnation of posterity.

Possessing a great talent for scenic display, Napoleon dressed his victories so as to get the fullest spectacular effect from them. He could pose now as the conqueror of Austria, and offset the *gloire* of his uncle's Marengo with that of his own Magenta. He had more *bâtons* and dukedoms to bestow, — more trophies to deposit in the Invalides. The gazettes, the official historians, the court writers, the spell-bound populace, acclaimed the new triumphs. Europe became too small for Imperial France to swagger in. Napoleon the First had meddled in Egypt, and Palestine, and the West Indies; his nephew must do likewise, and seek new worlds to conquer over sea.

Already, however, sober observers noted other symptoms, and soon the list of Imperial reverses grew ominously long. Early in 1860, Central Italy became a part of Victor Emanuel's kingdom: Napoleon had insisted that it should form a new state for his cousin Plon-Plon. That autumn, Sicily and Naples united themselves to Italy: Napoleon had wished and schemed otherwise. That same year, too, England compelled him to renounce his protectorate over Syria. Then he planned a French empire in Mexico; sent French troops over under Bazaine; set up Maximilian, who appeared to have grafted Napoleonism on our continent. But in 1867 he recalled his army, — "spontaneously" as he said. The world smiled when it reflected that the spontaneity of his withdrawal had been superinduced by a curt message from the United States and the massing of United States troops on the Rio Grande. In 1864 he would have kept Prussia and Austria from robbing Denmark; but as he had only words to risk, they heeded him not. In 1866, when Prussia and Austria went to war, expecting that Austria would be the victor, he had arranged to take a slice of Rhineland while Austria took Silesia. But Prussia was victorious, and so quickly that Napoleon could not save his reputation even as mediator.

At last Europe realized that his nod was not omnipotent, — that Prussia, his enemy, could raise herself to a power of the first rank, not only without but against his sanction. Napoleon also realized that his prestige was tottering. He must have some compensation for Prussia's aggrandizement. But when he asked for a strip of Rhineland, Bismarck replied: "I will never cede an inch of German soil." Napoleon, not ready for war, cast about for some other screen to his humiliation; for even in his legislature men now dared to taunt him with allowing Germany to grow perilously strong. To this taunt one of the Imperial spokesmen retorted, "Germany is divided into three fragments, which will never come together." A day or two later Bismarck published the secret treaties by which North and South Germany had bound themselves to support each other in case of attack.

Thus thwarted, Napoleon schemed to buy the tiny grandduchy of Luxemburg, which had long been garrisoned by Prussian troops. The King of Holland, who owned it, agreed to sell it for ninety million francs. Europe was willing, but Bismarck said *no*. He would consent to withdraw his troops, to destroy the fortifications, and to convert Luxemburg into a neutral state; more than that he would not allow. And with that Napoleon had to content himself, and to persuade the French — as best he could — that he had frightened the Prussians out of the grandduchy.

In 1863 Bismarck said to a friend: "From a distance, the French Empire seems to be something; near by, it is nothing." About the same time Napoleon, who had had much friendly intercourse with the Prussian statesman, said: "M. de Bismarck is not a serious man."

Just as the Luxemburg affair was concluded, all the world went to Paris to attend the Exposition, which was intended to be, and seemed, a symbol of the permanence of the Second Empire. The projectors knew that the immense preparations would enable the government to employ many workmen, who might otherwise be unruly, and that the vast concourse of visitors would bring money to the tradespeople and keep them from grumbling. The ostensible purpose, however, was to dazzle both Frenchmen and strangers by a view of Imperial magnificence; and it was fully achieved.

Paris herself, the Phryne among cities, astonished those who had never seen her, or who had seen her in old days. Where, they asked, were the narrow, crooked streets, in which barricaders once fortified themselves? Were these boulevards, stretching broad and straight, — were these they? And by what magic had the old, irregular dwellings been transformed into miles of tall, stately blocks? New churches, new quays, new parks, new palaces, bearing the impress of grace, symmetry, and a unifying planner, excited the wonder of the cosmopolitan throng of visitors. But the products of industry, the triumphs of the arts of peace, were not allowed to obscure the military glories of the Second Empire. A "Bridge of the Alma" and a "Boulevard of Sebastopol" kept the Crimean prowess in memory; a "Solferino Bridge" and a "Magenta Boulevard" bore witness to the Italian triumphs. And there were pageants, military, courtly, artistic; balls, at which, among the picked beauties of the world, the Empress Eugénie shone most beautiful; banquets, at which Napoleon sat at the head of the table, with monarchs at his right hand and his left deferentially listening. Little did the on-lookers suppose that the master of those magnificent revels had been lately frustrated by M. de Bismarck, who was merely one of the million whose presence in Paris seemed a tribute to Napoleon's supremacy.

History, it is said, never repeats: but is the

saying true? Is there not an old, old story of Belshazzar and the magnificent feast he gave in ancient Babylon, and the mysterious writing on the wall? And was not another Belshazzar repeating the episode in this modern Babylon less than thirty years ago? However that may be, the Exhibition of 1867 was the last triumph of Imperial France.

Imperialism had made a great show, reproducing, so far as it could, the glamour of the First Empire. Judge how potent that First Empire must have been, when mere imitation of it could thus hypnotize France and delude Europe! But Imperialism, generated by a crime and vitalized by corruption and deceit, was not *all* France. Honest France, excluded in the beginning, could not, would not, be lured in later. Napoleon would have conciliated, but the men whom he needed to conciliate would not even parley. To offset Victor Hugo and patriots of his rigid defiance, the Emperor had the outward acquiescence of Prosper Mérimée, the worldly courtier; of Alfred de Musset, the weak-willed, debauched poet; and of such as they. But he had the conscience of France against him; to offset *that* he leagued himself with Jesuits and Clericals. Having exhausted the expedients of force, he had tried the arts of flattery; he had intimidated, he had blandished;

he had made vice easy and attractive, in order that, though he could not win over the stubborn to his cause, their character might be softened through voluptuousness. Whosoever could be corrupted — let us give him full credit — he did corrupt in masterful fashion; but conscience, in France as elsewhere, is incorruptible.

Despite his complicated machinery for gagging conscience, protests began to be made boldly. One such protest, uttered towards the end of 1868, rang throughout France; and well it might, so audacious was the eloquence of the protester. Several newspapers had opened a subscription for a monument to Baudin, a Republican killed in the *coup d'état*. The proprietors of these newspapers were arrested. One of them, Delescluze, had for his advocate Léon Gambetta, a vehement young lawyer from the South. Before the judge, and the prosecuting attorneys, and the police — all myrmidons of the Emperor — he arraigned the Empire, closing with these words: " Here for seventeen years you have been absolute masters — 'masters at discretion,' it is your phrase — of France. Well, you have never dared to say, ' We will celebrate — we will include among the solemn festivals of France — the Second of December as a national anniversary! And yet all the governments which have succeeded each other in the land have honored the day of their birth; there are but two anniversaries — the 18th Brumaire and the 2d of December — which have never been put among the solemnities of origin, because you know that, if you dared to put these, the universal conscience would disavow them!" Gambetta's invective did not save his client from prison, but his arraignment of the Empire echoed throughout France.

And all the next year, 1869, though Imperialism abated in language none of its pretensions, it showed in deeds many signs of nervousness. No longer did it think it prudent, for instance, to abet the enormous extravagances of Hausmann, the remodeler of Paris. It even talked Liberalism, and set up a seeming Liberal Cabinet, with Ollivier at its head. "All the reform you may give us, we accept," said Gambetta bluntly; "and we may possibly force you to yield more than you intend; but all you give, and all we take, we shall simply use as a bridge to carry us over to another form of government." Evidently the conscience of France, expressing itself through the Republican spokesman, could not be placated or seduced.

A still blacker omen ushered in 1870. Pierre Bonaparte, the Emperor's cousin, shot in cold blood a journalist, Victor Noir. Two hundred thousand persons followed the victim's hearse; two hundred thousand voices shouted through the streets of

Paris, "Vengeance! Down with the Empire! Long live the Republic!" In April the ministers proposed further reforms, and called for another plebiscite, that worn-out Napoleonic device for deceiving public opinion. Seven and a third million votes were dutifully registered for the Empire, and only a million and a half against it; but the Imperialists did not exult, — a majority of voters in Paris, and forty-six thousand soldiers, had voted *no*.

To be deserted by the Parisians, on whom Napoleon had lavished so much pomp, — that, indeed, was hard; but the disaffection in the army meant danger. One desperate remedy remained, — a foreign war. Victory would bring to Imperialism sufficient prestige to postpone for several years the impending collapse; meanwhile, public attention would be diverted from grievances at home.

Nemesis saw to it that rogues thus minded should not lack opportunity. The Spaniards having elected an obscure German prince to be their king, the French ministers announced that they would never suffer him to reign. Of his own motion, the German prince declined the election, but the French were not appeased. They would humble the King of Prussia by forcing from him a meek promise. King William refused to be bullied; the French ministers proclaimed that Franco had been insulted. Not Imperialists only, but Frenchmen of all parties clamored for satisfaction. That love of *gloire*, that mercurial vanity which, twenty years before, had made them an easy prey to Louis Napoleon, now made them abettors of his breakneck venture. He appealed to their patriotism, the last refuge of a scoundrel, and they were beguiled.

War came, the Emperor being, by common report, most reluctant to consent to its declaration. He was its first victim. Five weeks after taking the field, he surrendered with nearly one hundred and ninety thousand men at Sédan. The corruption which through twenty years he had fostered, in all parts of the state where he expected to profit by it, had gangrened the army also, that branch which a military tyrant needs to have honestly administered. And now in his need the army failed him. He had been caught, as every one is caught who imagines that he can be wicked with impunity and still keep virtue for an ally when he needs her. From top to bottom his war department was rotten. Conscripts had, by bribe, evaded service; generals had sworn to false muster-rolls; ministers had connived with dishonest contractors. At Sédan, Napoleon paid the penalty of the corruption which he had erected into a system; at Sédan, moreover, he completed that cycle of parallels and imitations which he had made the business of his life. Just as Prussian Blücher paralyzed the last rally of the great Napoleon at Waterloo, so Prussian Moltke achieved the ignominy of Napoleon the Little at Sédan.

Men forget, even when they do not forgive. Frenchmen, furious at the humiliation of Sédan, cursed Napoleon as the author of it. But after a quarter of a century, although they have not forgiven him, they have come to look on him as victim rather than as villain. Later writers have held him up to be pitied. They describe his long years of suffering from the stone; they paint him during that month of August, 1870, as a poor, abject creature of circumstances, driven to bay by an irresistible foe, buffeted, scorned, despised by his own officers and troops. They show him to us, speechless and in agony, lifted from his horse at Saarbrücken; or huddled into a third-class railway carriage with a crowd of common soldiers escaping from the oncoming Prussians; or sitting, as cheerless as a death's-head, at a council of war; now lodged in mean quarters; now passing gloomily down regiments on their way to defeat, and never a voice to cry *Vive l'Empereur;* ever growing more and more haggard and nervous with worry, disaster, and endless cigarettes; continually pelted with telegrams from Empress Eugénie at Paris,

"Do this — do that, or the Empire is lost;" until that final early morning interview with Bismarck in the weaver's cottage at Donchéry. Latter-day Frenchmen, beholding such misery, have forgotten that Napoleon himself was chiefly responsible for it, and have ceased to execrate.

In closing, let us read, from a letter Bismarck wrote to his wife the day after the surrender, a description of the meeting of Napoleon and his conqueror: —

"*Vendresse*, Sept. 3, 1870. Yesterday morning at five o'clock, after I had been negotiating until one o'clock, A. M., with Moltke and the French generals about the capitulation to be concluded, I was awakened by General Reille, with whom I am acquainted, to tell me that Napoleon wished to speak with me. Unwashed and unbreakfasted, I rode towards Sédan, found the Emperor in an open carriage, with three aides-de-camp and three in attendance on horseback, halted on the road before Sédan. I dismounted, saluted him just as politely as at the Tuileries, and asked for his commands. He wished to see the King. I told him, as the truth was, that his Majesty had his quarters fifteen miles away, at the spot where I am now writing. In answer to Napoleon's question where he should go, I offered him, as I was not acquainted with the country, my own quarters at Donchéry, a small place in the neighborhood, close by Sédan. He accepted and drove, accompanied by his six Frenchmen, by me and by Carl (who in the mean time had ridden after me), through the lonely morning, towards our lines. Before reaching the spot, he began to be troubled on account of the possible crowd, and he asked me if he could alight in a lonely cottage by the wayside. I had it inspected by Carl, who brought word it was mean and dirty. '*N'importe*' (No matter), said N., and I ascended with him a rickety, narrow staircase. In an apartment ten feet square, with a deal-table and two rush-bottomed chairs, we sat for an hour; the others were below. A powerful contrast with our last meeting in the Tuileries in '67. Our conversation was difficult, if I wanted to avoid touching on topics which could not but affect painfully the man whom God's mighty hand had cast down. I had sent Carl to fetch officers from the town, and to beg Moltke to come."

That morning the terms of capitulation were drawn up, and the next day Napoleon went a prisoner to Wilhelmshöhe, whence, in due time, he was allowed to depart for England. At Chislehurst, on January 9, 1873, he died, having lived to see not only the extinction of French Imperialism and of the temporal Papacy, but also the creation of the German Empire and the union of Italy. To prevent all of these things had been his aim.

In a life like Garibaldi's we see what a disinterested genius can do by appealing to men's noble motives: the career of Louis Napoleon illustrates not less clearly what a man with talents and without scruples can accomplish by appealing to the instincts of vainglory and selfishness and terror; to the instinct which bullies weak nations and hoists the flag where it does not belong; to the instinct which has not the courage to acknowledge an error, but is quick to impute injuries, and declares that there shall be one conscience for politicians and another for citizens. Let us not flatter ourselves that only the French have cherished these stupendous delusions; let us rather take warning by the retribution exacted from them.

> "Forgetful is green earth: the Gods alone
> Remember everlastingly; they strike
> Remorselessly, and ever like for like.
> By their great memories the Gods are known."

KOSSUTH

The history of Hungary is in this respect unique: it records the career of an alien tribe which, cutting its way from Eastern Asia to the heart of Europe, founded there a nation, and this nation, after the friction of a thousand years, still preserves its racial characteristics. In 894 Duke Arpád led his horde of Magyars — whose earlier kinsmen were Huns and Avars — up the valley of the Danube. Long were they a terror to Europe; then, gradually, they had to content themselves with Hungary as their home. They became Christians; they adopted a monarchical government; alongside of their Aryan neighbors, they took on mediæval civilization. Europe, unable to expel or to destroy, acknowledged them as citizens. The time came when the Magyars, in a conflict lasting fivescore years, defended Europe against the invasion of another horde of Asiatic barbarians; till, unsupported by their neighbors, the Magyars succumbed to the Turks in the battle of Mohács in 1526. Afterwards, for one hundred and fifty years, Hungary herself writhed in the hands of the Mussulman; when that bondage ceased, she had a different oppressor, — Austria.

The Hungarian monarchy was elective, and after the battle of Mohács the Magyars chose for their king the sovereign of the Austrian states. The succession continued in the House of Hapsburg, becoming in fact hereditary; but, before the Magyars accepted him as king, each Hapsburg candidate must be ratified by the Hungarian Diet, and must swear to uphold the Hungarian Constitution. When, however, the expulsion of the Turks, at the end of the seventeenth century, left the Austrian sovereigns free to exercise their authority, they set about curtailing the ancient liberties of Hungary. Throughout the eighteenth century that process went on: the Magyars protested; the Emperor-King encroached, or, when the protests threatened to pass into insurrection, he paused for a while and gave fair promises.

Such was the situation when the French Revolution, followed by Napoleon's colossal ambition, startled Europe. During the quarter century of upheaval, the Magyars, still pouring their grievances into Vienna, remained loyal to their King. After Napoleon's downfall, the Old Régime being firmly reëstablished, Emperor Francis not only failed to keep his promises towards Hungary, but revived the old policy of Austrianization, which meant the substitution of German for Magyar officials, and the removal of the chief branches of government to Vienna. Again the protests became angry, until Francis, baffled and alarmed, convened the Diet. With the year 1825, when that Diet met, began the modern struggle of Hungary to recover that home rule which one after another of her Hapsburg kings had solemnly sworn to respect, and had as perfidiously disregarded. Thus the seed of the Magyar revolution was sown, like that of so many others, in a demand for the restoration of acknowledged rights, and not in a demand for innovation. Home rule, — Hungary to govern herself, instead of being bullied by foreigners who happened to be also subjects of her Emperor-King, — that seemed an object as simple and definite as it was just. Experience soon showed, however, that this cause was not simple; that it no more could be attained alone than gold can be taken from quartz without crushing the quartz and separating the silver and lead, and the crushed quartz itself, from the desired gold. For Hungary was imbedded in an old civilization, which must be broken up before home rule, and many another modern ideal, could be attained.

Imagine a country having an area about as large as the State of Colorado, inhabited by people sprung from four different races, — the Magyar, the Slav, the German, and the Italian: imagine, further, these races subdivided into eight different peoples, — Magyars, having poor kinsmen called Szeklers; Slavs, sending forth four different shoots, Slovaks in the North, Croats in the Southwest, Serbs in the South, and Wallachs in the East; imagine this motley population holding various creeds, — Roman Catholic, Greek Catholic, Calvinist, Lutheran, and Unitarian: imagine not merely each race, but each people, cherishing its own language, its own customs, its own ambitions, which inevitably clashed with those of its neighbors: and having imagined all this, you have not yet come to the end of Hungary's complex organism. Beside the conflicts of race and creed, there were political and social complications.

The dominant race was the Magyars, who numbered, however, only a third of the total population; their prevailing system was the feudal. A few hundred great nobles, or magnates, a considerable body of small nobles and a multitude of artisans, tradesmen, and peasants made up the social strata. Every Magyar who could trace descent to Arpád and his followers — though he were but a peasant in condition — was a noble: members of all the other races had no political rights. Hungary proper comprised fifty-two counties, each of which had its local congregation or assembly, which met four times a year, and sent suggestions or bills of grievances to the Central Diet, composed of the Table of Magnates and the Table of Deputies. A Palatine or Viceroy, representing the Sovereign, was the actual head of the kingdom. Outside of Hungary proper, the Croats had their local Diet at Agram, and Transylvania had hers; both also chose representatives to the Hungarian Diet. In a measure, therefore, we may call Hungary a federation, not forgetting, however, that it was a federation in which one race, the Magyars, domineered. The Latin language was the common medium of communication between Hungary and Austria, and among the diverse peoples.

The most significant event of the Diet of 1825 was the use by Count Stephen Széchenyi of the Magyar language instead of the Latin. Széchenyi, having traveled in Western Europe, came back imbued with large schemes of progress. He helped to introduce steamboats on the Danube; he founded a Magyar Academy; he proposed to join Buda and Pesth by a suspension bridge. By stimulating the material welfare of his country, he hoped that many of the social abuses would vanish without a struggle. And now his use of the Magyar language was a symptom of the awakening of the spirit of nationality, — one of the controlling motives in the history of Europe during the nineteenth century. In Hungary, as elsewhere, the arousing of that spirit was evidenced not only by an intenser political life, but also by a literary revival.

In direct reforms the Diet of 1825 accomplished little, — the Austrian government being still adroit in postponing a settlement, — but it was important in so far as it revealed the presence of new forces, whose nature was as yet undetermined. By the time another Diet assembled, in 1832, several questions had taken a definite shape. Foremost, of course, was Hungary's demand of home rule, in which all Magyars stood side by side; but when it came to internal affairs, they inevitably disagreed. The advanced Liberals proposed to emancipate the serfs, to extend the suffrage, and to abolish many of the privileges of the aristocracy. How grievous was the condition of the Hungarian serf may be inferred from the fact that, in spite of an improvement decreed by Maria Theresa, he was still bound to contribute to his landlord the equivalent of more than one hundred days' labor a year; he had no civic rights, and no other chance of redress than in the manorial court presided over by his master. The nobles, on the other hand, paid no taxes, ruled the county assemblies, appointed magistrates, and, except in case of a foreign invasion, rendered no military service, in return for all their exorbitant immunities.

That Magyar aristocracy has played so prominent a part in the history of Hungary that we may pause a moment to describe it. In 1830 the Magyar magnate was still the most picturesque noble in Europe. Like the Spanish grandee and the Venetian senator of an earlier time, he represented one of the highest expressions of the privileged classes. He was haughty, but warm-hearted; emotional, but brave: appeal to his honor, to his magnanimity, and — as Maria Theresa found — he would forget his grievances, disregard his interests, and devote himself body and soul to your cause. He might be ignorant, a spendthrift, an exacting master, but in his capacity for generosity he was — by whatever standard — truly a noble. In old times his forefathers had assembled every year, or when an emergency required, on the plain of Rákos, — a host of gallant warriors, in brilliant armor and gorgeous cloaks and trappings. There they deliberated — perhaps chose a king or deposed one — and then each rode home with his retinue, to live in a splendor half-barbaric for another year. In his dress the Magyar had an Oriental love of color, and in his music there is a similar glow, a similar charm.

As late as 1840 both the magnates and the

lesser nobility clung to their national costume as loyally as to their national constitution. "It now consists of the *attilla*," writes Paget at that date, "a frock coat, reaching nearly to the knee, with a military collar, and covered in front with gold lace; over this is generally worn, hanging loosely on one shoulder, the *mente*, a somewhat larger coat, lined with fur, and with a fur cape. It is generally suspended by some massive jeweled chain. The tight pantaloons and ankle-boots, with the never-failing spurs, form the lower part. The *kalpak*, or fur cap, is of innumerable forms, and ornamented by a feather fastened by a rich brooch. The white heron's plume, or aigrette, the rare product of the Southern Danube, is the most esteemed. The neck is opened, except for a black ribbon loosely passed round it, the ends of which are finished with gold fringe. The sabre is in the shape of the Turkish scimitar; indeed, richly ornamented Damascus blades, the spoils of some unsuccessful Moslem invasion, are very often worn, and are highly prized.

"The sword-belt is frequently a heavy gold chain, such as our ancient knights wore over their armor. The colors, as in many respects the form, of the Hungarian uniform, depend entirely on the taste of the individual, and vary from the simple blue dress of the hussar, with white cotton lace, to the rich stuffs, covered with pearls and diamonds, of the Prince Esterházy.

"On the whole, I know of no dress so handsome, so manly, and at the same time so convenient. It is only on gala days that gay and embroidered dresses are used; on ordinary occasions, as sittings of the Diet, county meetings, and others in which it is customary to wear uniform, dark colors with black silk lace, and trousers, or Hessian boots, are commonly used."[1]

Such, in its dress, was the Magyar aristocracy which the reformers set themselves to overcome; and in their character those Magyar nobles — were they magnates or simply gentlemen — cherished a tenacity of class unsurpassed by any other aristocrats in Europe. Nevertheless, the reformers boldly put forth a programme which involved the complete social and political reorganization of the country, — even throwing down a challenge to the aristocracy to surrender privileges in which these deemed their very existence rooted. Parties had begun to array themselves on these lines when Louis Kossuth entered public life.

Born at the village of Monok, Zemplen County, on April 27, 1802, Kossuth had for his father a lesser noble, Slavic in origin, Lutheran in faith,

[1] John Paget, *Hungary and Transylvania* (new edition, New York, 1850), i, 249, 250.

and lawyer by profession. The son received a good education, and began to practice law, which led easily to politics. He sat in his county assembly, was early conspicuous as an advocate of popular rights and as an eloquent speaker. Thus equipped, he took his seat in the Diet of 1832, where, as proxy to a magnate, he had a voice but no vote. There seemed slight chance of his emerging from his proxy's obscurity, but to genius all conditions are fluid. Kossuth conceived the plan of publishing the reports of the debates in the Diet. The government permitted no newspapers, and trimmed all other publications to suit its views; but the members of both Houses could speak freely, without danger of arrest for any of their utterances in the Diet. To circulate their speeches would, therefore, as Kossuth saw, put within reach of the Hungarians a mass of political reading not otherwise obtainable. Hardly had he begun to publish, ere government signified its desire of buying his press. Deprived of this, he employed secretaries who wrote out his abstracts of the proceedings and sent them through the mails to their destination. Government ordered its postmen to confiscate and destroy. Still unvanquished, Kossuth dispatched his budgets by special messengers. Government was foiled. By these devices, before the close of the Diet in 1836,

Kossuth — the obscure magnate's proxy — had become one of the most widely known men in the kingdom. The reports were literally *his* reports, giving not only the tenor of the chief debates, but also his comments thereon.

He now proposed to edit in similar fashion the proceedings of the quarterly meetings of the fifty-two county assemblies; but Government, no longer restrained by his inviolability as member of the Diet, arrested him. He spent two years in prison, denied books and all intercourse with his friends, before his case came to trial: then he was sentenced to a further confinement of four years, during which his great solace was the study of Shakespeare.

Meanwhile, political and social agitation was swelling. The King, thinking a European war over the Eastern Question imminent, summoned another Diet to vote him a fresh subsidy and more soldiers. But the Diet, indignant and headstrong, refused all help till Kossuth and some other political prisoners should be released. The King yielded. Kossuth came forth a national hero.

After several months spent in recuperating his health, Kossuth, in January, 1841, established the *Pesti Hirlap*, or *Pesth Gazette*. That Government acquiesced in this project showed how far the tide of Liberalism had risen. It showed, too, that Government was astute,—hoping in this way to rob Kossuth of his martyr's halo; deeming it wiser to let him publish openly than surreptitiously; trusting, above all, to the sharpness of its censors' eyes and scissors. Kossuth, on his side, was equally cunning, versed in the art of dressing his opinions in such guise that the censor could not object to them, though they carried a meaning which his readers knew how to interpret according to his intention. He wrote on all topics with a vehemence and an Oriental heat which won him tens of thousands of admirers. Like any Magyar patriot, he could count on one of the most powerful of allies,—the race hatred between his countrymen and the Austrians. The very word "German" signified, in the Magyar language, *vile, base, despicable.* There was a Magyar proverb to the effect that "German is the only language God does not understand." Innumerable illustrations of this antipathy might be cited, but the following, which Paget tells, will serve as well as another: The proprietor of a theatre produced what he considered a fine piece of scenery, in which was represented a full moon, with round, fat, clean-shaved face. When it rose, the audience hissed, and shouted, " Down with the German moon ! " The manager took the hint; next night there rose a swarthy-cheeked, black-moustachioed orb. Hurrahs burst from every mouth, and all cried, "Long live our own true Magyar moon!"

Doubt not that Kossuth knew how to kindle the fuel which ages of hatred had been storing. He had the gift peculiar to really great popular leaders of appealing directly to racial pride and passion; so it mattered little that he dealt in generalizations. Speaking broadly, he preached the abolition of feudalism and the aggrandizement of the Magyar nationality. The former purpose brought him and the Liberals into conflict with the conservative aristocracy; the latter inflamed against the Magyars the long-smouldering hatred of their subject peoples.

For the spirit of nationality had awakened these also. The Slavs of Croatia, Slavonia, and Dalmatia dreamed of establishing a great Slavic kingdom in Southeastern Europe; they, too, were putting forth a literature. Their Illyrism — to their prospective nation they gave the name "Illyria" — clashed with the recrudescent Magyarism. When the Hungarian Diet decreed that the Magyar language should be taught in their schools, and that every official must use it, they protested as strenuously as the Magyars themselves had protested when Austria tried to impose the German language and German officials on them. "The Magyars are an island in the Slavic ocean," exclaimed Gaj, the poet and spokesman of Illyrism, to the Hungarian Diet: " I did not make this ocean, I did not stir up its waves; but take care that they do not go over your heads and drown you." Nevertheless, the law was passed. In the Southland the Serbs along the Danube, in the East the Wallachs of Transylvania, feeling the first tingle of national aspirations, resented this encroachment. Austria — whose motto was, *Divide et impera* — found her advantage in embittering tribe with tribe and class with class.

For three years and a half Kossuth's *Gazette* had an unprecedented influence in Hungary; but in the summer of 1844, disagreeing with his publisher over a matter of salary, he resigned, and expected to found another journal which should draw off the *Gazette's* patrons. Government, however, refused to grant him a license. Accordingly, he devoted himself to agitation in another form. In the assembly of the County of Pesth, he discussed with matchless eloquence the great political questions; outside, he organized an economical crusade. Austria burdened Hungary with a tariff which stunted her industrial and commercial development. Kossuth created a league whose members vowed for five years to use only Hungarian products. He projected a railway to Fiume, to secure an outlet for exporting Hungarian goods. He urged the establishment of savings banks and of mercantile corporations. And for a brief time, under this patriotic stimulus, trade flourished.

Thus through all the arteries of the body politic new blood was throbbing. Give a people a great idea, and they will find how to apply it to every concern of life. The Magyar Liberals were surely undermining feudalism; their race was growing more and more restive at Austria's obstinate delays. When Austria removed the native county sheriffs and put German administrators in their stead, all the Magyar factions joined in denouncing such an assault on their national life. The county system had been the safeguard of Hungary's political institutions for well-nigh eight hundred years; the sheriff was the foremost official in the county, to whose guidance its interests and civic activity were intrusted. To make an alien sheriff was therefore to check national agitation at its source. Accordingly, the Diet which met in the autumn of 1847 met full of defiance and resentment, though the platform of the Liberals, drawn up by the judicial Deák, wore on its surface a conciliatory aspect. After a hot canvass, Kossuth was elected to represent Pesth in the Chamber of Deputies. A few sessions sufficed to establish his preëminence as an orator, and his leadership of the Liberal party.

During the winter months of 1847–48 but little was done, though much was discussed. As usual, the Magnates resisted the reforms aimed at their class; as usual, Government temporized and postponed. Suddenly, at the beginning of March, 1848, news reached Presburg of the revolution in Paris, and of the flight of Louis Philippe. That news passed like a torch throughout Europe, kindling as it passed the fires of revolt. At Presburg, on March 3, Kossuth rose in the Diet and interrupted a debate on the financial difficulties with Austria. That question of finance, he said, could never be settled separately; in it was involved the whole question of Austria's disregard of Hungary's rights. Hungary must have her own laws, her own ministry; taxation must be equal; the franchise must be extended. More than that, he added, Hungary could never prosper until every part of the Empire should be governed by uniform constitutional methods.

Kossuth's "baptismal speech of the revolution" took the Lower House by storm. An address to the Throne was framed, which, after fruitless reluctance on the part of the Magnates, a large committee, headed by Kossuth and Count Louis Batthyányi, — the Liberal leader in the Upper House, — carried twelve days later to Vienna. The delegates found the Austrian capital in an uproar.

On March 13 Metternich, deserted by the aristocracy on whose behalf he had labored unscrupulously for fifty years, had been hounded from office. The people, after a bloody struggle, had possession of the city, and they welcomed Kossuth as a deliverer; for his " baptismal speech " had made their aims articulate.

The next day, Emperor Ferdinand received the deputation very graciously, and promised to grant their petition. Exulting, they returned to Presburg. A Cabinet was formed in which Batthyányi held the premiership, and Kossuth the portfolio of finance. Soon, very soon, tremendous difficulties beset them: Radicals clamored for a republic; the subject races revolted; the Imperial government proved perfidious.

The key to Austria's subsequent conduct is this: Austria, at heart a coward, had long been able to play the bully; now, however, her outraged peoples had risen in wrath and held her at their mercy; the bully cringed, promised, conceded; concession brought a temporary respite from danger; thereupon she began to think she had been unduly terrified and to regret her concessions; so she cautiously put out feelers of arrogance, to resume her rôle of bully. When she met sharp resistance, she quickly drew back again, to await a better opportunity. Throughout this crisis, Emperor Ferdinand, at his best a man of mediocre capacity, was becoming imbecile through epilepsy, and a Court clique, or Camarilla, ruled him and the Empire.

All this was not yet clear to the Hungarians. Assuming the Imperial assurances to be honest, they passed a reform bill abolishing the privileges of the nobles, who were to be compensated by the state for the loss they sustained in the emancipation of their serfs. Bills authorizing equal taxation, trial by jury, freedom of speech, the abolition of tithes, and the extension of the franchise to one million two hundred thousand voters, were adopted with but little discussion. Religious toleration — except for Jews — became the law of the land.

The Magnates having made this unparalleled sacrifice, King Ferdinand came over to Presburg and dissolved the Diet in a speech approving its action, and reiterating his pledge to uphold the Constitution. The Cabinet proceeded to organize its administration, — a task which would have been sufficient at any time to keep it busy, but now extraordinary and urgent matters pressed upon it. The Wallachs, Serbs, and Croats rose in rebellion. Most alarming was the situation in Croatia, where the Slavs were agitating for separation from Hungary. Baron Jellachich, who had just been appointed Ban or Viceroy of Croatia, abetting the insurrection, strengthened the Croat army. In June the Magyar ministers hurried to Innspruck — whither the Emperor and Camarilla had fled after a second outbreak in Vienna — to protest against these rebellious acts. The Emperor assured them that he had given the Ban no sanction; that he had, indeed, dismissed him from the Imperial service. It happened that Jellachich was at Innspruck at this very moment, carrying the notification of dismissal in his pocket, and in his mind an unwritten commission to serve Austria against Hungary.

The rebellion of the Serbs, accompanied by unspeakable atrocities, was openly fomented by Austrian agents; likewise the outbreak in Transylvania. Hungary's embarrassments increased; she had still to accept Ferdinand's assurances of good faith, for he was her legal king; but now she knew that the Camarilla, the actual Imperial government, was instigating her enemies.

The newly elected National Assembly convened at Pesth, the ancient capital, early in July. The royal address condemned by implication Jellachich and all rebels, but the insurrection grew in violence from day to day. On July 11 Kossuth made in the Assembly the most effective speech of his life. Posterity stands incredulous before the record of great orators who, Orpheus-like, are said to have moved stocks and stones by their voice; yet not on this account must we disbelieve the record. For posterity can never supply the one thing needful to the consummate orator's success, — it can never supply the state of mind of his audience. We shall always find that the epoch-making speech was addressed to listeners every one of whom had long been burning to hear just those words. This is why so many of the orations that altered history look faded on the printed page; this is why we must in many cases judge the orator as we judge the singer or the actor, — by the effect he produces on his contemporaries. Kossuth, by this standard, ranks with the first orators of the century, though a later generation is little thrilled by his printed speeches. Men who heard him, even those who heard him speak in a language not his own, and who had listened to Webster and Clay and Choate, declare that they never heard his equal. Upon his own countrymen, to whom his words came charged with the associations which belong to one's mother-tongue, his eloquence was irresistible.

In that 11th of July speech, at least, we, too, after long years, can feel the glow. The occasion itself was dramatic. Every deputy realized that the crisis of the revolution was at hand, — that

Hungary must either turn back, or dare to plunge into an unknown and perilous sea. All were waiting for the decisive word.

Kossuth, just risen from a bed of sickness, with tottering steps mounted the tribune. He was a man of medium height; his hair was brown, his eyes blue; he wore a full mustache and cut his beard sailor-wise, so that it formed a shaggy fringe beneath his smooth-shaven chin. At first, as he spoke, his pallid face and feeble gestures, though they enhanced the solemnity of his words, made his hearers dread a collapse; but presently he seemed to be fired with the strength which burned in his subject, and they listened for two hours, spell-bound and electrified.

"I feel," he said to them, "as if God had put in my hands the trumpet to rouse the dead, that, if sinners and weak, they may sink back into death, but that, if the vigor of life is still in them, they may waken to eternity." He then went on to review the quarrel with Croatia, declaring that to that country Hungary had, from immemorial time, accorded all the privileges which she herself enjoyed, and that recently she had conceded to the Croats a wider use of their native language. "I can understand a people," he said ironically, "who, deeming the freedom they possess too little, take up arms to acquire more, though they play, indeed, a hazardous game, for such weapons are two-edged; but I cannot understand a people who say, 'The freedom you offer us is too great, — we will not accept your offer, but will go and submit ourselves to the yoke of Absolutism.'" Kossuth next touched on the situation in the South, and showed wherein it differed from that in the Southwest. He told how the Camarilla had sought to compel the ministers to acknowledge the unlawful pretensions of Croatia, and thereby to annul the pledges of the King. He pointed out, as an ominous cloud on the eastern horizon, the recent appearance of a Russian army along the Pruth. When, after this review, he solemnly announced, "The fatherland is in danger," not a deputy was surprised, not a head shook incredulously. At last he asked for authority to levy two hundred thousand soldiers, and to raise a loan of forty-two million florins, setting forth the means by which he planned to meet this extraordinary measure as eloquently as he had set forth its need.

He had held the Assembly captivated for two hours; now, as he was closing, his strength failed, and he could not speak. The deputies, too, were speechless. For a brief moment intense silence reigned between him and them. Then Paul Nyáry, who only yesterday had attacked the policy of the Cabinet, rose, lifted his right hand as if invoking

God to be his witness, and exclaimed, " We grant everything!" In a flash four hundred hands were raised, and four hundred voices repeated Nyáry's covenant. When quiet came again, Kossuth had recovered strength to say that his request should not be taken as a demand for a vote of confidence. " We ask your vote for the preservation of the country; and, sirs, if any breast sighs for freedom, if any desire waits for fulfilment, let that breast suffer a little longer, let it have patience until we have saved the fatherland. You have all risen to a man, and I bow before the great-heartedness of the nation, while I ask one thing more: let your energy equal your patriotism, and the gates of hell itself shall not prevail against Hungary!"

In March, under the magic of Kossuth's irresistible oratory, the Magyars had boldly demanded their constitutional rights; now in July, thrilled by the same magic, they pledged themselves to defend their independence to the death.

The summer passed amid recruiting of Honvéds, volunteer "defenders of the fatherland," the attempt to quell the insurrection in Transylvania and among the Serbs, and the renewed intrigues of the Imperial Court to browbeat the Hungarian Cabinet. In September, Jellachich, at last avowedly in the service of Austria, prepared to invade Hungary.

The Palatine, unable to bring about a reconciliation, quitted the country. The Viennese Cabinet appointed Count Lamberg to assume full control of the military affairs in the kingdom; the Hungarians pronounced his appointment unconstitutional, and they were right. On his arrival at Pesth, he was murdered by a mob. This rash crime caused some of the Liberals to withdraw horrified. Batthyányi resigned the premiership, and a Committee of National Defense, in which Kossuth predominated, was chosen. On October 2, the Camarilla, grown truculent, dispatched Recsey to dissolve the Hungarian Assembly, and bade Hungary to submit to Jellachich. The Magyars heeded neither command. Having equity and law on their side, they acted henceforth on the assumption that the orders which emanated from Vienna could not be attributed to Ferdinand without imputing perjury to him.

War could no longer be avoided. The Committee of National Defense displayed great energy in organizing resistance. Kossuth's eloquence went over the land, and the cloddish peasant left the plough, the well-to-do tradesman deserted his shop, the lawyer dropped his brief, to become volunteers in the service of their country. A third outbreak at Vienna sent the Camarilla hurrying off to Olmütz, and seemed for a moment to assure the final triumph of the revolution. During the three weeks which elapsed before an Austrian army under Prince Windischgrätz — he who said that "human beings begin with barons" — could be brought up, the Hungarians debated whether they should go to the assistance of the Viennese, for they wished to be strictly legal. At last they found justification in the plea that they had a right to pursue Jellachich, who was marching to join Windischgrätz, across the Austrian frontier. But they decided too late. Their troops were beaten at Schwechat, on the outskirts of Vienna, just as Windischgrätz was successfully storming the city (October 29).

For six weeks thereafter Windischgrätz devoted himself to stamping out the rebellion in Vienna, and in preparing for a campaign against Hungary. On December 2 poor, weak-witted Ferdinand abdicated, and his nephew Francis Joseph succeeded him as emperor. This change betokened the returning confidence of the Court party. They now felt sure of crushing the revolution, and of restoring the Old Régime; but they had no intention that, when the rest of Austria was re-subjected to their despotism, Hungary alone should enjoy a constitutional government. Yet this had been promised by Ferdinand, and he had scruples against openly violating his oath. Therefore, by remov ing him and substituting Francis Joseph, they had a sovereign unhampered by pledges. To this scheme the Magyars naturally did not bend; their Constitution was their life, and that Constitution recognized no king who had not been crowned by the Magyars, and had not sworn to preserve their rights inviolate.

Ten days before Christmas, Windischgrätz opened his campaign. Five armies besides his own invaded Hungary from five different directions. The Magyars had employed the six weeks' lull in defensive preparations. They gave Arthur Görgei, an ex-officer thirty-one years old, — able, stern, selfish, and inordinately ambitious, — the command of the Army of the Upper Danube. He proposed to abandon the frontier and to mass the Hungarian forces in the interior, where they could choose their own ground; but the Committee of Defense insisted that every inch of Hungarian soil should be contested. A fortnight's operations proved the wisdom of Görgei's plan: the Magyars were easily driven back, and on New Year's eve the Austrians camped within gunshot of Buda-Pesth. The following day, January 1, 1849, a melancholy procession of ministers, deputies, state officials, fearful citizens, and stragglers, set out from Pesth, carrying with them the precious crown of St. Stephen, the public coffers and archives, and the printing-presses for bank-notes. Debreczin, a town forty leagues inland, became the temporary capital. At Buda-Pesth, Windischgrätz celebrated his triumph by holding a Bloody Assize. To envoys from the fugitive government who asked him to state his conditions, he only replied, " I do not treat with rebels."

Among the Magyars, consternation was quickly succeeded by a mood of desperation, — such a mood as made France invincible in 1792. Again did Kossuth's eloquence pass like the breath of life over the land; again did his energy direct the equipment of new recruits and fill the gaps of the regiments already in the field. Had the deputies at Debreczin voted as they wished, they would have voted for peace; but they knew that the majority of their countrymen would reject any peace which Austria was likely to offer, and they were ashamed to appear less daring than Kossuth.

The enthusiasm, we might call it the recklessness, with which the Magyars rallied to repel invasion, became a people who counted John Hunyádi and Francis Rakóczy among their national heroes. Thanks to their patriotic fervor, the Hungarian cause, which seemed about to collapse at the beginning of January, seemed about to prevail at the end of March. Bem had worsted the Wallachs and Austrians in Transylvania; Görgei had redeemed Northern Hungary; Klapka and Damjanics had brought Windischgrätz to bay in the midlands.

Well had it been for Hungary if these astonishing successes had prevented internal discord, for twofold dissensions now threatened to sap the growing strength. From one side, the generals chafed at being subordinate to the civilian Committee of Defense; on the other, a large body of soldiers and of civilians were angry at the evident drift of Kossuth and his friends towards a republic. Görgei, the most conspicuous of the generals, led this opposition. He declared in a manifesto that the army would fight to maintain against every foreign enemy the Constitution granted by Ferdinand, but that they would favor no attempt to convert the constitutional monarchy into a republic. The Committee of Defense, most eager in their patriotism, could not refrain from meddling; they suffered from the delusion common to such committees, and believed that they knew better than the trained men of war how war should be waged. They felt, too, political responsibilities which made them all the more active; and they had, as was natural, their favorites among the officers. Had the government been strong, it would have cashiered Görgei; being weak, and solicitous of conciliating so important a man, it tolerated him.

But when a government and its generals distrust each other,— as we learned in our civil war,— conciliation can satisfy neither. If Görgei lost a battle, his enemies charged him with lukewarmness or disobedience; he retorted by blaming the committee for failing to support him or for breaking in upon his plans. We need not sift the recriminations in detail: it suffices for us to know that, from January on, Görgei and Kossuth, and their respective partisans, worked thus at odds.

Nevertheless, among the masses these quarrels had but slight effect. The average Magyar was simply bent on avenging his long score of oppression against Austria. He realized that his own existence depended on that of Hungary, and to him Kossuth's eloquence was like a trumpet-call of duty. That in performing his duty the Magyar might lawfully wreak vengeance on his oppressors, made duty doubly attractive.

In the early spring, Austria closed the way to compromise by proclaiming a new charter for the whole Empire. This charter declared that all the provinces of the Empire were to be reduced to a common equality, deprived of local rights, and governed by a central administration at Vienna.

The Magyars, then, had nothing to hope. Whether they submitted to Austria or were conquered by her, their ancient Constitution would be blotted out. They would cease to be a nation. Accordingly, on April 14, 1849, they proclaimed the independence of Hungary, calling God and man to witness the wrongs she had suffered from the House of Hapsburg, and setting forth the illegality, truculence, and perfidy of Austria during the past thirteen months. A diet was to be summoned, which should determine the form of government that Hungary would permanently adopt; meanwhile Kossuth was chosen president-governor, and by appointing Görgei commander-in-chief he hoped to heal old wounds.

The moment was propitious. The Austrians had been beaten in a great battle (at Isaszeg) on April 7; and most of the fortresses, except Buda, had been recaptured. Görgei himself seemed satisfied. The elated Magyars dreamed even of a swift campaign against Vienna, and of bringing the Imperial tyrant to terms which should be acceptable to all his subject races. But their dream, if ever attainable, was spoiled by delay. Görgei insisted that Buda must be retaken before he marched farther west, and only on May 21 did he succeed in storming its citadel. By that time a new peril, more terrible than any previous, loomed up. Austria, in despair of subjugating Hungary, had besought Russia to help her, and the Czar, glad of an excuse for interfering, was marshaling his troops on the Hungarian frontier.

No assistance could the Magyars secure to offset this threatened intervention. France and England would not even recognize their republic, although Frenchmen and Englishmen privately sympathized with their cause. From Venice alone, the little republic round whose neck the Austrian noose was already tightening, came a heartfelt recognition, which, however, added not a soldier to their army nor a florin to their purse. Desperate, but not yet willing to surrender, the Magyars nerved themselves for a final effort. Kossuth proclaimed a crusade, a levy in mass; every man to arm himself, were it only with a scythe or a bludgeon; perpetual prayers to be offered up in the churches; the enemy to be harassed at all places, to be hindered by the destruction of bridges and stores, and, wherever possible, by open fighting.

Posterity, calmly reviewing a death struggle like this, is amazed that any people could be roused to make that last stand. Plainly enough, the Magyars had three soldiers against them to every one of theirs; ammunition and victuals were failing them; their treasury was empty; their armies could expect no reinforcements: to what end, therefore, protract a hopeless war? Reasoning thus, we miss the secret, not only of the revolutionists of 1848–49, but of all who have ever been kindled by patriotism to defend a cause they held dearer than life. The Magyars would never have gone thus far, — never have felt during that May-month the fleeting exhilaration of victory, — had they not been fired by a passion which not disaster but death alone could quench.

The Russian invasion being assured, the Magyar government held a council of war, at which it was proposed to consolidate the various armies, and to defeat first the Austrians coming from the west and then the Russians coming from the north and east, — a sensible plan, frustrated, however, by delays, some of which were unavoidable. The Austrian army, strengthened by reinforcements from Italy, and commanded by Marshal Haynau, who came red-handed from Brescia, advanced into Hungary, and defeated Görgei on the river Waag (June 20-21). The Magyar Government and Diet departed for the second time, in melancholy procession, from their capital. By the middle of July one hundred and fifty thousand Russians — eighty thousand of whom were led by the wolfish Paskevitch — had penetrated into the heart of the country. Inevitably, the Magyar forces would be driven in and caught between the victorious enemies: nevertheless, they would not yet submit.

Internal discord alone tarnished the record of the last days of the Hungarian Republic. On

July 1, Kossuth removed Görgei for insubordination, but Görgei's officers and men protested so loudly that Kossuth thought it discreet to reinstate him. Three weeks later, a fraction of the Diet, assembled at Szegedin, declared the equality of all the races in Hungary, emancipated the Jews, and then, warned by the rumble of hostile cannon, it dissolved forever.

For yet a few weeks we have news of Kossuth hurrying hither and thither to proclaim hope where no hope was; conferring with nonplussed but still resolute generals; dragging after him, like his shadow, those printing-presses for banknotes, now worth no more than blank paper. Finally, at Arád, he resigned the presidency, and appointed Görgei dictator with full powers. At Világos, on August 13, Görgei surrendered his exhausted army of twenty-three thousand men to Rüdiger, the Russian general. Thus was consummated what the Magyars, frenzied by defeat, branded as Görgei's treason, but what, to an impartial observer, appears an inevitable act. Görgei's course throughout the war cannot be commended: inordinate personal ambition, not treason, was its motive; he may have thought to play the part of Monk, but more likely he had taken Napoleon for his model; one thing alone is certain, — he did not intend that Kossuth should reap the glory of victory, if victory came. In surrendering at Világos he did what every commander is justified in doing, when further resistance could only entail fresh losses without any hope of altering the result.

Learning the capitulation of the main army, the other generals one by one submitted. Klapka alone maintained an heroic defense at Comorn until September 27, when hunger and an empty magazine forced him to surrender. With the hauling down of the red-white-and-green flag from the citadel of Comorn vanished the last symbol of that revolution which, bursting forth at Palermo in January, 1848, had spread through Europe, shaking the thrones of monarchs, and kindling in down-trodden people the belief that a new epoch, a Golden Age of Liberty, had come. Hopes as splendid as men ever cherished had now been shattered, and in their stead only the bitterest memories remained; for as each people pondered in sorrow and oppression the events of those twenty months, it was tormented by the reflection that its own dissensions, not less than the might of its enemies, had wrought its ruin.

Austria, careful by a deceitful silence to encourage the stray bodies of Magyar troops to give themselves up, proceeded to punish Hungary with a severity which matched the persecutions of the

French Reign of Terror. In every city Marshal Haynau set up his shambles; in every parish he plied his scourge. Imprisonment, torture, confiscation, overtook the lowly defenders of the Magyar cause; death awaited the leaders. On October 6, at Arád, fourteen generals were hanged or shot, and that same day Count Louis Batthyányi was shot at Pesth. Görgei was spared, thanks to the personal intervention of Czar Nicholas.

Kossuth and several thousand Magyars took refuge in Turkey. The Sultan protected him, in spite of the threats of Russia and Austria, — protected him because the Turkish religion forbade the betrayal of a refugee, — but kept him for nearly two years in half bondage. Then the Magyar hero, at the instance of the American Congress, was permitted to embark on an American man-of-war. He came to the United States, where he was greeted with an enthusiasm which no other foreigner except Lafayette had stirred. He got boundless sympathy, and no inconsiderable sum of money for prosecuting the emancipation of Hungary; but the times were unfavorable, and the lot of the Magyars concerned very little the rulers of European diplomacy after 1850. Returning to Europe, Kossuth made agitation his sole aim. He strove to interest the great powers in Hungary's fate; he strove, through secret emissaries, to provoke the Magyars themselves to rebel. The former were deaf; the latter, taught by terrible experience, deemed it folly to attack Austria again in the field. Through the persistent and judicious political agitation led by the sagacious Francis Deák, they achieved, in 1867, a recognition of their constitutional rights, and a full measure of home rule.

Kossuth, however, refused to the last to be reconciled. He lived in exile at Turin, a forlorn old man, forlorn but inflexible, amid the memories of exploits which once had amazed the world. There he died on March 20, 1894, having survived all his contemporaries, friends and foes alike, who had beheld the rise and splendor and eclipse of his astonishing career. To be the mouthpiece of a haughty and valiant people at one of the heroic crises of their history was his mission. His genius, his defects, mirror the genius and defects of his countrymen; his glory, being a part of the glory of a whole race, is secure. That race, which Arpád led into the heart of Europe, showed, at Kossuth's summons, a thousand years later, that it had not lost the traits which had once distinguished it on the shores of Lake Baikal and along the upper waters of the Yenisei.

GARIBALDI

When men look back, two or three hundred years hence, upon the nineteenth century, it may well be that they will discern its salient characteristic to have been, not scientific, not inventive, as we popularly suppose, but *romantic.* Science will soon bury our present heaps of facts under larger accumulations, from the summit of which broader theories may be scanned; to-morrow will make to-day's wonderful invention old-fashioned and insufficient: but the romance with which this later time has been charged will exercise an increasing fascination over poets and novelists and historians, as the years roll on. Oblivion swallows up material achievements, but great deeds never grow old. That many of our writers should not have heard this note of the age argues that they, rather than the age, are prosaic and commonplace. For to what other period shall we turn for a richer store of those vicissitudes and contrasts in fortune which make up the real romance, the profound tragedy, of life? Everywhere the dissolution of a society rooted in mediæval traditions is accompanied by confusion and struggle, — the birth-pangs of a new order. Classes whose separation seemed permanent are thrown together, and antagonistic elements are strangely mixed; there is strife, and doubt, and excess; sudden combinations are suddenly rent by discords; anachronisms flourish side by side with innovations; new institutions wear old names, and old abuses mask in new disguises.

In such a crisis, two facts are prominent: the unusual range of activity offered to the individual — may he not traverse the whole scale of experience? — and the dependence of the individual upon himself. He rises, or he falls, by his own motion. The privileges of caste avail nothing; for the very confusion produces a certain wild equality, whereby all start at the line, and the swiftest wins. Napoleon's maxim, *La carrière ouverte aux talents*, is the motto of the century. Napoleon himself is an epochal illustration of the power of the individual to make the momentum of circumstances work for him. The Revolution, it is true, had harnessed the steeds; but Napoleon dared to mount the chariot, grasped the reins, and drove over Europe, upsetting thrones and princedoms and hierarchies. The haughty descendants of immemorial lineage gave place to the brothers and comrades of the " Corsican upstart." Murat, the son of a tavern-keeper; Ney, a briefless law-student; Lannes, a dyer; Soult, Masséna, Berthier, Junot, soldiers of fortune; and how many other children of the Third Estate, — laughed at the pretensions of humbled Bourbons, Hapsburgs, and Hohenzollerns! Frequent reactions between revolution and restoration serve to emphasize the stress of this crisis; and these contrasts in the conditions of men, revealing human character under the most diverse phases, show how inextricably the romantic and the tragic are interwoven in the average lot.

Nor in Europe only has this spectacle been going forward. The United States also have witnessed similarly rapid transmutations, partly due to other causes. Within a generation we have seen a gigantic national upheaval: three millions of artisans, clerks, merchants, and lawyers were transformed by the magic of a drum-beat into soldiers; and then, the conflict over, soldiers and uniforms vanished, and the labors of peace were resumed.

Follow Abraham Lincoln from his Illinois log-cabin to the White House; follow Grant from his tanyard to Appomattox, — and you can compute the sweep of Fortune's wheel. These careers were lived so near us that they hardly astonish us; they seem as natural as daylight; and in truth they are as natural as that or any other every-day miracle. As if forgetful of these, we ransack the past, or fiction, or melodramas, for heroes to admire. To weak imaginations, distance still lends enchantment.

Our age has produced one romantic man, however, who had not to wait for the mellowing effects of time to be recognized as romantic. He enjoyed, almost from the outset of his career, the fame of a legendary hero, and he will, we cannot doubt, be a hero to posterity. Some future Tasso will find in his life a theme nobler than Godfrey's, too romantic in fact for either invention or myth to enhance it. He lived dramas as naturally as Shakespeare wrote them; the commonplace could not befall him. Looking at him from one side we might say, "Here is a Homeric hero, strangely transplanted from the *Iliad* into an era of railroads and telegraphs!" But if we fix our attention on other qualities, we discover in him a typical democrat, fit product of a democratic age. This man was Joseph Garibaldi, whose career alone would suffice to redeem the nineteenth century from the stigma of egotism and the rebuke of commonplaceness.

Among all the political achievements of our century, none has more of noble charm than the redemption of Italy. Whether we look at the difficulty of the undertaking, or at the careers of the leaders and the temper of the people who engaged in it, we are alike allured and amazed. After the fall of the Roman Empire, Italy had never been united under one government; nevertheless, from the time of Dante on, the aspiration towards national unity was kept burning in every patriotic Italian heart. During the Middle Age, little republics won independence by overthrowing their feudal lords; then they quarreled among themselves; and then they became the prey and appanage of a few strong families. The Bishop of Rome, forgetful of his spiritual mission, lusted after worldly power, established himself as a temporal sovereign, and elevated his cardinals into temporal princes. Foreign invaders — Normans, Spaniards, Germans, French — swept over the peninsula in successive waves; bloodshed and pillage signalized their coming, corruption was the slime they left behind them. One by one, the refugees of independence were submerged in the flood of servitude; until at last Venice herself, become merely the mummy of a republic, crumbled to dust at Napoleon's touch. Napoleon promised, but did not give, to Italy the unity or the freedom which she still dreamed of: he parceled her anew into duchies and kingdoms. By that act he broke down ancient barriers and opened a new prospect. Italians beheld the old order, which had so long oppressed them that many believed it must endure perpetually, suddenly dissolved, and in its stead a change, although not the change they longed for. Still, any change, in such circumstances, implies fresh possibilities; and the Italians passed from a lethargy which had seemed hopelessly enthralling into a restless wakefulness.

The twenty years of the reign of Force, of which Napoleon was the embodiment, ended at Waterloo. Europe, exhausted, sank back into conservatism, and was ruled for thirty years by Craft, of which Metternich was the symbol.[1] The Congress of Vienna reimposed the past upon Italy. Monarchs and bureaucrats, like children who amuse themselves by " making believe " things are not as they are, would have it appear that the deluge of revolution, with all its mighty wrecks and subversions, had never been. The Pope was restored in the States of the Church; the Bourbons ruled again in Naples and Sicily; an Austrian was Archduke of Tuscany; Parma and Piacenza were assigned to Napoleon's wife, Maria Louisa; Venetia and Lombardy went as spoils to Austria; an absolutist king reigned in Piedmont. Evidently the revolution had been but a summer thunderstorm,

---

[1] After Metternich, we have the period of Sham-Force, under Louis Napoleon; and finally of Force again, under Bismarck. These four stages complete the cycle of European politics during the past century.

for the sun of despotism was shining once more. The sun shone; but what of the sultry air? What of the threatening clouds along the horizon? Were these the fringe of the dispersing storm, or the portents of another? Mutterings and rumblings, too, Carbonari plottings, and quickly extinguished flashes of insurrection, — did not these omens belie Diplomacy's pretense that the eighteenth century had been happily resuscitated to exist forever?

It was during this interval of reaction and relapse, when hope was stifled and energy slept; when victorious despotism flattered itself with the belief that the Napoleonic episode had demonstrated the absurdity of Liberalism; when Metternich, the spider of Schönbrunn, was spinning his cobwebs of chicane across the path to liberty, — then it was that the generation which should live to see Italy free and united was getting what learning it could in the Jesuit-ridden schools. Of this generation the most romantic figure was Giuseppe Garibaldi.

Joseph Garibaldi was born at Nice, July 4, 1807. His father was a fisherman, thrifty enough to have a small vessel of his own. Such stories as have come down to us of the boy's childhood show him to have been plucky, adventurous, and tenderhearted. He cried bitterly at having broken a grasshopper's leg; he rescued, when only seven, a laundress from drowning; he sailed off with some truant companions for Genoa, and might have vanished forever, had he not been overtaken near Monaco and brought home. His education was intrusted to two priests, from whom, he says, he learned nothing; then to a layman, Arena, who gave him a smattering of reading, arithmetic, and history. As he was quick at learning, his parents wished to make a lawyer or a priest of him; but he had the rover's instinct and could not resist the enticements of the sea. At length, when he was fourteen, his parents yielded, and he became a sailor.

Of those early voyages, we need mention only one, which took him to Rome. Immense the impression the Holy City made on his imagination! He saw not the Rome of the Cæsars, nor the Rome of the Popes, — the city whose monuments entomb twenty-five centuries of history; but, he says, "the Rome of the future, that Rome of which I have never despaired, — shipwrecked, at the point of death, buried in the depth of American forests; the Rome of the regenerating idea of a great people; the dominating idea of whatever Past or Present could inspire in me, as it has been through all my life. Oh, Rome became then dear to me above all earthly existences. I adored her with all the fervor of my soul. In short, Rome for me is

Italy, and I see no Italy possible save in the union, compact or federate, of her scattered members. Rome is the symbol of united Italy, under whatever form you will. And the most infernal work of the Papacy was that of keeping her morally and materially divided." [1]

Thenceforth the young mariner, who rose rapidly to be mate and master, could not rest for the thought of the Eternal City, and of the country his patriotism craved. During these years, he learned to take Fortune's buffets: he was captured by pirates, he lay ill and penniless for months at Constantinople, — adventures which in another career would demand more than passing notice, but which he deemed unimportant in comparison with a conversation he had with a young Ligurian, who unfolded to him the dreams of the Mazzinians. "Columbus did not experience so great a satisfaction at the discovery of America," says Garibaldi, "as I experienced at finding one who busied himself with the redemption of our fatherland."

Fatherland! the name seemed a mockery to the Italians of that time. Italy, as Metternich phrased it, was only a geographical expression. Seven or eight petty princes, including the Pope, ruled the little patches into which the Peninsula was cut

[1] This was written in 1849.

up. All the north, except Piedmont, was directly subject to a foreign despot, Austria; while, indirectly, Austria domineered over Tuscany, Rome, and Naples. Piedmont had a native king, indeed, but Absolutism throve nowhere more vigorously than there. The Jesuits controlled the worship and education of the little kingdom; reactionaries filled the ministerial offices, the army, and the government bureaux; the sovereign himself, Charles Albert, believed devoutly in the divine right of kings, and held that it would be criminal in him, by granting his people more freedom, to lessen the responsibility imposed on him by God. Throughout the Peninsula, no one might discuss politics, whether in speech or writing. It was high treason to suggest representative government; the sovereign's will was the only constitution. In some parts of the land, the very word *Italy* could not be used by actors on the stage; and everywhere censors kept watch to prevent the idea of a regenerate Italy from slipping into print.

By foreigners, the Italians were more often despised than pitied; they were believed to be pluckless, wordy, deceitful creatures, who at best had their uses as singers, dancing-masters, and painters' models. Among themselves, discord (born of ancestral feuds), envy (born of local ambitions, a love of haranguing, and a lack of leaders), had thrice resulted in an abortive revolution. And now, just as the third attempt had failed, and in its failure had discredited the great organizations of conspiracy that had been for fifteen years the hope of Italian patriotism, Joseph Mazzini, a Genoese a year younger than Garibaldi, banished from Piedmont because he had a suspicious habit of walking abroad after dark, formed the new secret society of Young Italy which aimed at not only the political but the social and moral redemption of his countrymen. Garibaldi, eager to hasten the emancipation of his country, joined Young Italy; but in the first plot in which he was engaged his confederates failed to appear at the appointed time, and he was forced to fly from Genoa for his life. "Here begins my public career," he says in his memoirs.

After being twice captured and twice escaping, he made his way on foot, disguised as a peasant, to Marseilles, where, on opening a newspaper, the first thing he read was the sentence of death decreed against him should he ever be caught in Piedmont. This was in February, 1834. Proscribed but not disheartened, when chance offered he resumed his seafaring. But mercantile voyages grew monotonous. Should he offer his services to the Bey of Tunis, who was seeking a European to take charge of his navy? After hesitation, Garibaldi decided "no." During a cholera epidemic, he volunteered as nurse in the Marseilles hospital. Finally he shipped for South America. Landing at Rio Janeiro, he fell in with another exile, Rossetti, and for a while they kept a shop. Soon, however, more congenial occupation presented itself.

Rio Grande do Sul, the southernmost province of Brazil, had revolted from the Empire and set up a republic, which it was struggling to maintain. Garibaldi, who could never resist aiding republicans, equipped a small privateer, on which he and Rossetti, with twelve companions, set sail for the south. This was the opening of a life of adventure which lasted twelve years, and which, could we trace it step by step, would be found a nonpareil of heroic deeds and startling dangers. The political and social condition of South America then resembled in lawlessness that period in European history when chivalry had its rise; when, as a foil to the bullying and craft and greed of the many, stood out the courage and honor and courtesy of the few. Garibaldi, whether by sea or land, approved himself a peerless knight. Following him, we should witness now a battle of gunboats far up the river Parana, until, his ammunition having given out, he loaded the cannon with the chain cables; or, again, we should undergo the horrors of a shipwreck near the mouth of La Plata, or join in a desperate battle against great odds at some lonely Paraguayan ranch; we should traverse vast pampas, or thrid the solitude of trackless forests; we should know hunger, thirst, and cold, and be incessantly attacking or attacked; and we should realize that although these campaigns seem mere border forays when compared with the wars of modern Europe and the United States, yet they settled the fate of territories as large as France, and required those martial qualities which beget heroism in any crisis under any sky.

Although we must pass all this, one marking episode in Garibaldi's life at that time ought not to be forgotten. His ships had been cast away in a storm. He succeeded in swimming to shore, but his dearest comrades perished. He felt lonely, dispirited, and though he was soon to command another cruiser, the excitements of war could no longer dissipate his melancholy. "In short," he says, in a characteristic passage of his Autobiography, "I had need of a human being to love me immediately, — to have one near without whom existence was growing intolerable to me. Although not old, I understood men well enough to know how hard it is to find a true friend. A woman? Yes, a woman; for I always deemed her the most perfect of creatures, and — whatever may be said — amongst women it is infinitely easier to find a loving heart. I was pacing the quarter-deck, ruminating my dismal thoughts, and, after reasonings of all kinds, I decided finally to seek a woman, to draw me out of my tiresome and unbearable condition. I cast a casual glance towards the Barra: that was the name of a rather high hill at the entrance of the lagune, toward the south, on which were visible some simple and picturesque habitations. There, with the aid of the glass, I discovered a young woman. I had myself set ashore in her direction. I disembarked, and, going towards the house where was the object of my expedition, I had not reached her before I met a man of the place, whom I had known at the beginning of our stay. He asked me to take coffee in his house. We entered, and the first person who met my gaze was she whose appearance had caused me to come ashore. It was Anita, the mother of my sons, the companion of my life in good and evil fortune, — the woman whose courage I have so often envied. We both remained rapt and speechless, reciprocally looking at each other, like two persons who do not meet for the first time, and who seek in the features one of the other something to assist recollection. At last I greeted her and said, 'Thou must be mine.' I spoke but little Portuguese, and uttered these hardy words in Italian. However, I was magnetic in my presumption. I had drawn a knot, sealed a compact, which death alone could break."

A few nights later Garibaldi carried Anita off to his ship, clandestinely as it appears, and they were wedded when they reached another port. She was a companion matching his ideal: she shared his wild fortunes and hardships; she was an indefatigable horsewoman, a dead-shot, and upon occasion she could touch off a cannon.

After years of fighting, Garibaldi obtained a furlough, gathered a drove of cattle, and journeyed across Uruguay to Montevideo. There he was reduced to teach the rudiments of arithmetic in a private school, picking up whatever other precarious pennies he could, until civil war broke out in Uruguay, and he enlisted on the side of the people, struggling to free themselves from a bloodthirsty dictator. Garibaldi's exploits as a guerrilla and corsair had made him famous, and now he repeated at Montevideo his amazing feats. From among his countrymen he organized an "Italian Legion," which proved throughout a long service that Italians could and would fight, — two facts which scornful Europe was loth then to believe. He also illustrated his perfect disinterestedness by refusing all rewards beyond a bare means of subsistence. At a time when he held the fate of Montevideo in his hand, he had not money to buy candles to light the poor room where he and his family were dwelling.

Thus, giving his utmost for liberty and the welfare of strangers, he saw the years pass without bringing the one thing he desired most of all, — the chance to consecrate himself to the redemption of Italy. That desire, the ruling passion of his life, had followed him everywhere. I marvel that any materialists exist; for where, in the material world, shall we find anything comparable to the tenacity of ideas? Think not to preserve them by locking them in an iron safe; write them not on stone, which crumbles, but on the human soul, and they shall be indestructible. Have we not daily proof that against remorse, love, hate, ambition, all the powers of the material world — fire or frost, hunger, disease, persecution — dash as harmless as vapor against adamant? By the moral precepts, by which Moses awed his people three thousand years ago, we are awed. They are permanent, being graven on something more durable than tables of stone; and it matters not how many times old Nile is renewed, or whether Sinai itself wear in dust away.

On Garibaldi's heart of hearts "Italy" was written, — an ideal which nothing could cancel.

At length, in the early autumn of 1846 news came to Montevideo that a Liberal Pope had been elected at Rome, that the word "amnesty" had been uttered, and that the Peninsula was throbbing with splendid hopes. Each succeeding message confirmed the presentiment that the longed-for day of action was nigh. Garibaldi, subordinating his hatred of priestcraft to his patriotism, wrote to offer his sword to the new Pope, to whom all Italians were looking as the leader of their crusade for freedom, but Pius never acknowledged the offer. Then Garibaldi and some threescore of the Legion hired a brigantine, which they named *La Speranza* (Hope), and on April 15, 1848, bade the Montevideans farewell. They had to touch at Santa Pola, on the coast of Spain, for water, where they learned that all Europe was in revolution, and then they dropped anchor at Nice on June 23. Over Garibaldi's head the death-sentence still hung, but he had nothing to fear, as the events of the past six months had wiped out old memories. Those six months had had no parallel in modern European history. They had witnessed the triumph of revolution from the Douro to the Don.

Not even during the Napoleonic upheaval had modern Europe felt a convulsion like that of 1848: for government and order were as necessary to Napoleon as to his victims, and his revolution was the effort of one lion to devour foxes and wolves, — of one preponderant tyranny to absorb many smaller tyrannies; but the catastrophe of 1848 seemed, to anxious observers, to endanger civilization itself. Society was dissolving into its elements. The many-headed beast had risen, ubiquitous, terrific. Lop off one head, and others grew from the trunk. What substitute could possibly be found in that chaos for the tottering system? Nothing seemed certain but anarchy.

That was the year when sovereigns were suddenly made acquainted with their lackeys' staircases and the back doors of their palaces. The Pope escaped from Rome in the livery of a footman. Ferdinand, Emperor of Austria, fled twice from Vienna. Louis Philippe, the "citizen king" of the French, put on a disguise, and slipped away to England. Metternich, rudely interrupted in his diplomatic game of chess, barely escaped with his life to London. The Crown Prince of Prussia, subsequently Emperor of Germany, eluded the angry Berliners, a trusty noble driving the carriage in which he escaped. There was a scampering of petty German princes, as of prairie-dogs at the sportsman's approach. Nobility, whose ambition hitherto had been to display itself, was now wondrously fond of burrows. And just as the frightened upholders of absolutism went into hiding, the apostles of democracy emerged from prisons and exile.

Paper constitutions, grandiloquent manifestoes, patriotic resolutions, doctrinaire pamphlets, were whirled hither and thither as thick as autumn leaves. Every man who had a tongue spoke; speaking, so furious was the din, soon loudened into shouting. But the Old Régime was encamped in no Jericho whose walls would tumble at mere sound. There must be deeds as well as words; in truth, more action and less Babel had been wiser. Committees of national safety, working-men's unions, civic guards, armies of the people, sprang into existence, and it is wonderful to note with what quickness officers and leaders were found to command them. Universities were turned into recruiting stations and barracks; students and professors became soldiers. There were heroic combats, excesses, reverses bravely borne. Gradually the fatal lack of centre and organization could not be concealed. The leaders disputed as to measures; then followed misunderstandings, jealousies, desertions. Each doctrinaire cared that *his* plan, rather than the general cause, should prevail. Each sect, each race, feared that it would lose should its rival take the lead. But the purpose of monarchy was everywhere the same, — to recover its footing; and the agents of monarchy, cautiously creeping out of their retreats, began to profit by the divisions among their enemies. Within a year the European revolt was crushed. Nevertheless its lessons abide. It taught that despots cannot be permanently abolished so long as a large majority of a nation require despotic government, and the proof that they require it is the fact that they submit to it; whence it follows that real democracy cannot conquer until a people be educated up to the capacity of governing themselves. It taught that without unity among the heads and obedience among the members no reform can succeed. It taught, finally, that no society which has once attained a certain level of civilization can exist in a state of anarchy; for when anarchy is reached, the opportunity of the strongest man, the tyrant, offers, and the process of reconstruction from the basis of absolutism begins.

To Liberals, in June, 1848, however, the days of tyranny seemed at an end; the Golden Age of liberty, constitutional government, and the brotherhood of nations seemed to have dawned. Garibaldi learned that Lombardy had expelled the Austrians; that Charles Albert, the Piedmontese King, had drawn his sword as the champion of Italian independence; and that the Pope and the other princes, including even Bomba of Naples, had espoused the national cause. The rapid victories of the spring had been succeeded by military inertia; frantic enthusiasm had given place to a chatter of criticism; but not even those who grumbled loudest believed as yet that the cause was in danger.

Garibaldi hurried to the King's headquarters, near Mantua. He was no lover of royalty, but he would support any king honestly fighting in behalf of Italy. Charles Albert granted him an audience, but avoided accepting his offered services, telling him that he had better consult the Minister of War, at Turin. To Turin, accordingly, Garibaldi posted back, saw that official, received further evasive replies, and departed angry. To have traveled seven thousand miles over sea to fight for his country's redemption, only to be treated in this fashion, might well astound a blunt soldier who had supposed that every volunteer would be welcomed. In his own case the rebuff was peculiarly astonishing, for he was, presumably, an ally whom any commander would be glad to secure. Europe had rung with the fame of his South American career, and already regarded him as a legendary hero. Imagine Charlemagne refusing Roland's aid in his campaign against the Paynims, or the old Romans turning coldly away from one of the great Twin Brethren!

Although Garibaldi would have despised reasons of state which deprived him of the right of volunteering against Austria, yet the King had to be governed by them. For his excuse in declaring war had been that, unless he interfered, anarchy, followed by a republic, would prevail in Lombardy. To be consistent, therefore, he had to keep clear of even an apparent league with republicanism as embodied in Garibaldi.

Baffled and exasperated, but determined not to be cut off from all activity, Garibaldi went to Milan, where a provisional government with republican leanings still ruled. By it he and his legionaries were hospitably received, and sent out, with a considerable body of raw recruits, to harass the Austrians along the lakes. In a few weeks, however, the main Austrian army had reconquered Lombardy, and the Garibaldians were driven to take refuge across the Swiss frontier.

Garibaldi, like a true knight-errant, now went forth in search of another chance to do battle for freedom. At Florence the republicans did him honor, but were wary of asking him to command their troops, the fact being that each district had leaders of its own, and a host of zealous aspirants, who were patriotically disinclined to make way for even the most distinguished knight-errant. At Rome, whence Pius IX had fled, the revolutionists gave him a warmer greeting, and when, in February, 1849, they set up a republic, — Garibaldi having made a motion to that effect in the Roman Assembly, — they made him second in command of their army. And now, properly speaking, the tale of Garibaldi's European exploits begins.

We cannot follow in detail the story of the defense of Rome against the French troops sent thither by the perfidious Louis Napoleon, and their allies from Spain and Naples; yet it were well worth our while to give an hour to deeds so brilliant, so noble, so picturesque, — to pass from the Assembly Hall, where Mazzini, the indomitable dreamer, was the dictator, to the fortifications where band after band of volunteers, speaking many dialects, clothed in many costumes, were resolved to give their lives for freedom! We should see Lucian Manara, a modern knight, captain of a legion of brave men; we should see Mameli, the blond poet-soldier, a mere lad; and the brothers Dandolo, and Medici and Nino Bixio, and many another doomed to win renown by an early death there, or there to begin a career which became a necessary strand in Italy's regeneration. But, most conspicuous of all, we should see Garibaldi, for whom the legionaries and their leaders had such a feeling as the Knights of the Round Table for Arthur their King. Call it loyalty, 't is not enough; call it filial affection, something remains unexpressed; call it fascination, enthusiasm, sorcery, — each term helps the definition, though none singly suffices. His was, indeed, that eldest sorcery which binds the hearts of men to their hero, — that power which reveals itself as an ideal stronger than danger or hardship or disease, something to worship, to love, to die for.

During her five-and-twenty centuries, Rome had seen many strange captains, but none more original than this, her latest defender, from the pampas of South America. In person he was of middle stature; his hair and beard were of a brown inclining to red; his eyes blue, more noteworthy for their expression than for their color; his mouth, so far as it could be seen under the moustache, was firm, but capable of an irresistible smile. His soldiers, remembering his aspect in battle, spoke of his face as " leonine ; " women, caught perhaps by the charm rather than the cut of his features, thought him beautiful. And as if Nature had not done enough to mark her hero, he adopted on his return to Europe the dress which he had worn in South America, — a small, plumed cap, the grayish-white cloak or *poncho* lined with red, the red flannel shirt, the trousers and boots of the Uruguayan herdsmen and guerrillas.

During that siege of Rome, Europe came to

know Garibaldi and his red-shirted companions, who were equally bizarre in character and in costume, — a troop of poets, students, dreamers, vagabonds, and adventurers, — who, with the nucleus of the Legion from Montevideo, were capable, under their chieftain's guidance, of splendid achievements. Their victories against the Neapolitans at Palestrina and Velletri; their stubborn defense of Rome against the overwhelming armies of France; their bravery at the Villa Pamfili; their desperate struggle to hold the Vascello, where Manara was killed; their unwilling but inevitable yielding of the outposts, and finally of the inner breastworks, — made up a tale of heroism which could be matched only at Venice in that year of waning revolution.

But Europe had declared that there should be no republic at Rome, and after nine weeks' gallantry the city capitulated to the French, who represented the cause of reaction. Garibaldi, however, did not surrender. On the day when the French made their entry by one gate, he marched out of another, followed by nearly four thousand soldiers. He wound across the Campagna, and then for twenty-nine days he led his troop among the Apennines, evading now the French who pressed on the rear, now the Austrians, who harassed both flanks and threatened to bar the advance. The little army dwindled, but Garibaldi held his purpose to reach Venice, where the Austrian tyrants had not yet forced their return. At length, however, in the little republic of San Marino he was surrounded. All but two hundred of his followers disbanded; with the remainder he eluded the enemy's cordon, reached the coast at Cesenatico, seized some fishing-boats, and embarked for Venice. Mid-voyage, a fleet of Austrian cruisers came upon them and opened fire. As best they could the fugitives landed, with Austrian pursuers at their heels. Garibaldi and one companion bore Anita in dying condition — she had followed the retreat on horseback all the way from Rome — to a wood-cutter's hut, where she died. A moment later Garibaldi had to fly.

Of that retreat, and his subsequent hair-breadth escapes in being smuggled across Italy, he has left in his memoirs a thrilling account. For a second time he tasted the bitterness of exile: his first refuge was Genoa, but the Piedmontese government, timid after defeat, informed him that he must depart; he was expelled from Turin at the instigation of the French; England warned him that he must be gone from Gibraltar within a week. Only in semi-savage Morocco did he at last find shelter; thence, after a few months, he came to New York. Consider who it was that

Europe thus outlawed, and what was his crime. He was a man whose life had been a long devotion to human liberty, and whose most recent guilt was to have attempted to prevent foreign despots from reënslaving his countrymen. A system is judged by the men it persecutes.

Wifeless, homeless, chagrined by the thought that Italy had waged her war of independence only to be beaten, Garibaldi began his second wanderings. A real *Odyssey* we may call it, with its strange happenings. For a year the hero of Rome earned a bare livelihood making candles in Meucci's factory on Staten Island; then he shipped for Central and South America; captained a cargo of guano from Lima to Canton, and a cargo of tea back to Lima; brought a ship laden with copper, round Cape Horn to Boston; and finally, in May, 1854, he dropped anchor at Genoa, where the government no longer feared his presence. With the proceeds of his mercantile ventures, he bought Caprera,—a mere rock, which juts out of the Tuscan Sea, near the northern tip of Sardinia. There, "like some tired eagle on a crag remote," he dwelt five years, apparently oblivious to the passing current of events, and wholly intent on coaxing a few vines and vegetables to grow on his wind-swept rock.

Early in 1859 a messenger summoned Garibaldi

from his hermitage to Turin. This summons was not unexpected. For months the world had regarded war in Italy as inevitable, and now war was on the point of breaking out.

How had this come to pass? After her defeat in 1849, Piedmont, the little northwestern kingdom of four million souls, had sturdily set about reforming herself. She stood firmly by the constitutional government adopted in 1848; she strengthened her army and her navy; she took education out of the hands of the Jesuits; she encouraged commerce, industry, and agriculture. Thus she proved to Europe that Italians could govern themselves by as good a political system as then existed; to all the other Italians, groaning under Austrian, or Bourbon, or Papal tyranny, she proved that they might look to her to lead the Italian cause.

This marvelous attainment was due primarily to Count Cavour, the statesman who, since 1850, had been almost continuously prime minister of Piedmont; and, in the second place, to Victor Emanuel, the shrewd, honest, chivalrous King, worthy to be the visible symbol of Italy's patriotism. But Cavour had realized from the beginning that, however strong he might make Piedmont, she would not be able singly to cope with Austria: four millions against thirty-five millions — the odds were too great! So he labored to bring Piedmont into the stream of European life; he allied her to France and England in the Crimean War; and now, at the beginning of 1859 he had persuaded Napoleon III to march the armies of France into Italy to join Piedmont in expelling the Austrians.

All this had been brought about against great hindrances, not the least of which was the keeping in check the Italian conspirators. Since the days of the Carbonari, a certain number of Italians had hoped to set up a republic. Mazzini, now the chief leader of conspiracy, was uncompromisingly republican, holding so little faith in the methods of Cavour and the Constitutional Monarchists that he never hesitated to hatch plots against them as well as against the Austrians. Between these two irreconcilable parties Garibaldi was the link. By preference a republican, he yet recognized Victor Emanuel as the only practicable standard-bearer, and he therefore fought loyally under him; but he distrusted Cavour, scorned diplomacy, and abhorred Napoleon III. In his exuberant way, he insisted that Italians could, if they would, recover independence without begging the rogue, who had crushed Rome ten years before, to succor them.

A volunteer corps, called the Hunters of the Alps, was accordingly organized, with the double purpose of using Garibaldi's skill as a guerrilla chieftain against the Austrians, and his unique popularity in drawing all sorts of partisans to support the national war. He suspected that the government was not wholly ingenuous; he complained that his volunteers had to swallow many snubs from the regulars; he chafed at being responsible to any superior: but the fact that he had at last a chance of striking the oppressors of Italy outweighed everything else.

Despite the shortness of the war of 1859, Garibaldi and his Hunters proved of real service in it. Varese, Como, remember their valor still; and had not Napoleon III suspended hostilities after the great victory of Solferino, the Garibaldians might have redeemed the Tyrol. But Napoleon's peace of Villafranca, while it gave Lombardy to Piedmont, left Venetia in the hands of the Austrians, and stopped further operations in the north at that time. During the autumn, however, Garibaldi, with many of his volunteers, went to Tuscany, where a provisional government was then awaiting the propitious moment for annexation to Victor Emanuel's kingdom. The situation was very ticklish, requiring careful diplomacy: Garibaldi, who shared with General Fanti the military command, wished to have done with diplomacy, to call out one hundred thousand volunteers, and to rely on them to disentangle all complications.

Irritated at having his plan overruled, he resigned his command and withdrew to Caprera.

Within three months, however, he was called from his retreat. Secret agents brought word that "something could be done" in Sicily, where for a long time Mazzinians had been preparing a revolt. It needed, they said, but Garibaldi's presence to redeem the island from Bourbon misrule. He could not resist the temptation. Trusty lieutenants of his had collected arms and ammunition, hired two steamers and enrolled volunteers. At Genoa, where these preparations were making, nobody, except the government officials, was ignorant of their purpose. The government, however, pretended not to see. Cavour could not openly abet an expedition against a power with which Piedmont was not at war; neither did he wish to hinder an expedition for whose success he and all Italian patriots prayed. So he discreetly closed his eyes.

On the night of May 5, 1860, Garibaldi and 1067 followers embarked on their two steamers near Genoa and vanished into the darkness. For a week thereafter Europe wondered whither they were bound, — whether against the Papal States or Naples; then the telegraph reported that they had landed at Marsala, on the morning of May 11, just in time to escape two Neapolitan cruisers which had been watching for them. From that moment, day by day, with increasing astonishment, the world followed the progress of Garibaldi and his Thousand. No achievement like theirs has been chronicled in many centuries. They set out, a thousand filibusters, scantily equipped and undrilled, to free an island of two and a half million inhabitants, an island guarded by an army fifty thousand strong, with forts and garrisons in all its ports, and having quick communication with Naples, where the Bourbon King had six million more subjects from whom to recruit his forces. Grant that the Sicilians fervently sympathized with Garibaldi, yet they were too wary to commit themselves before they had indications that he would win; grant that the Bourbon troops were half-hearted and ludicrously superstitious, — many of them believed that the Garibaldians were wizards, bullet-proof, — yet they had been trained to fight, they were well-armed, and by their numbers alone were formidable. That they would run away could not be assumed by the little band of liberators, any more than Childe Roland could suppose that the grim monsters who threatened his advance would vanish when he upon his slug-horn blew.

And in truth the Bourbon soldiers did not run. At Calatifimi the Garibaldians beat them only after a fierce encounter; at Palermo there was a desperate struggle; at Milazzo, a resistance which might, if prolonged, have destroyed the expedition. In every instance it seemed as if the Bourbons might have won had they but displayed a little more nerve, another half hour's persistence; but it was always the Garibaldians who had the precious reserve of pluck and strength to draw upon, and they always won. Their capture of Palermo, a walled city of two hundred thousand inhabitants, defended by many regiments on land and by men-of-war in the harbor, ranks highest among their exploits. Less than a month after quitting Genoa, they had liberated more than half the island and had set up a provisional government. By the first of August only two or three fortresses had not surrendered to them.

And now questions of diplomacy came in to disturb the swift current of conquest. Garibaldi determined to cross to the mainland, redeem Naples, march on to Rome, and from the Capitol hail Victor Emanuel King of Italy. Cavour saw great danger in this plan. At any moment, a defeat would jeopard the positions already gained; an attack on the Pope's domain would bring Louis Napoleon and Austria to his rescue, and might entail a war in which the just-formed Kingdom of Italy would be broken up; furthermore, Cavour believed that assimilation ought to keep pace with annexation. He knew that it would require long training to raise the Italians of the south, corrupted by ages of hideous misrule, to the level of their northern kinsmen.

Such considerations as these could not, however, deter Garibaldi. He grew wroth at the thought that any foreigner — were he even the Emperor of the French — should be consulted by Italians in the achievement of their independence. Eluding both the Neapolitan and the Piedmontese cruisers, he crossed to the mainland and took Reggio after a sharp fight. From that moment his progress towards the capital resembled a triumph. And when, on September 7, accompanied by only a few officers, he entered Naples, though there were still a dozen or more Bourbon regiments in garrison there, the soldiers joined with the civilians and the loud-throated *lazzaroni* in acclaiming him their deliverer. Yet only a few hours before their King had sneaked off, too craven to defend himself, too much detested to be defended. Think what it meant that this should happen, — that the sovereign, the source of honor, the fountain of justice, the symbol of the life and integrity of the state, should not find in his own palace one loyal sword unsheathed in his defense, even though the loyalty were hired, like that of the eight hundred Swiss who gave their lives for Louis XVI! By an inevitable penalty, Bourbon misrule in Naples passed vilely away; it had been, as Gladstone declared, the embodied "negation of God:" even in its collapse and ruin there was nothing tragic, portending strength; there was only the negative energy of putrefaction.

Having taken measures for temporarily governing Naples, Garibaldi prepared for a last encounter with the Bourbons. King Francis still commanded an army of forty thousand men along the Volturno, near Capua. There Garibaldi, with hardly a third of that number, fought and won a pitched battle on October 1. A month later he welcomed Victor Emanuel as sovereign of the kingdom which he and his Thousand had liberated. The republicans, instigated by Mazzini, had wished to postpone, if they could not prevent, annexation; but Garibaldi, whose patriotic instinct was truer than their partisanship, insisted that Naples and Sicily should be united to the Kingdom of Italy under the House of Savoy. In all modern history there is no parallel to his bestowal of his conquests on the King, as there is nothing nobler than his complete disinterestedness. He declined all honors, titles, stipends, and offices for himself, and departed, almost secretly, from Naples for Caprera the day after he had consigned the government to its new lord.

Fortune has one gift which she begrudges even

to her darlings: she does not allow them to die at the summit of their career. Either too soon for their country's good, or too late for their personal fame, she sends death to dispatch them. Pericles, Cavour, Lincoln, were snatched away prematurely; Themistocles and Grant should have prayed to be released before they had slipped below their zenith. So, too, Garibaldi lacked nothing but that, after having redeemed a kingdom by one of the most splendid expeditions in history, and after having given it to the unifier of his fatherland, he should have vanished from the earth. Thanks to a kindlier fortune, the old Hebrew prophets were translated, and the Homeric heroes were borne off invisible, at the perfect moment. But while Garibaldi lacked this epic finale to his epic career, the closing decades of his life were as characteristic as any.

In the spring of 1861 he reappeared on the scene at the opening of the first parliament of the Kingdom of Italy, to which he had been chosen deputy by many districts. He came, not jubilant but angry. Nice, his home, had been ceded to France in payment for French aid in the war of 1859: against Cavour, who had consented to this bargain, Garibaldi conceived the most intense hatred, and on the floor of the House he fulminated at the Prime Minister whose "treason had made

Garibaldi a foreigner in his native land." He complained, further, because the officers and soldiers of the Garibaldian army had not been generously treated by the government. The outburst was most deplorable. Many feared that the hero's testiness might lead to civil war; and though the King arranged a meeting, in the hope of bringing about a reconciliation, Garibaldi went from it with bitterness in his heart. Six weeks later, on June 6, Cavour, stricken by fever, died when his country needed him most. Little did Garibaldi realize that in the great statesman's death he was losing the man who had been indispensable to his success in Sicily, and whose judgment was needed to direct Garibaldian impulses to a fruitful end.

Only Rome and Venetia now remained ununited to the Kingdom of Italy: in Rome a French garrison propped the Pope's despised temporal power; in Venetia the Austrian regiments held fast. To rescue the Italians still in bondage, and to complete the unification of Italy, were henceforth Garibaldi's aims. He paid no heed to the diplomatic embarrassments which his schemes might create; for as usual he regarded diplomacy as a device by which cowards, knaves, and traitors thwarted the desires of patriots.

In the summer of 1862, therefore, he recruited three or four thousand volunteers in Sicily, raised the war-cry, "Rome or death," crossed to the mainland, and had to be forcibly stopped by royal troops at Aspromonte. In the brief skirmish he was wounded, and for many months was confined at Varignano, whither flocked admirers — men, women, and youths — from all parts of Europe. There is no doubt that Rattazzi, then the premier, had connived at the expedition, hoping to repeat Cavour's master-stroke; but the conditions were different from those of 1860, and the Premier but illustrated the truth that talent cannot even copy genius judiciously. Moreover, by allowing Garibaldi to go so far and by then arresting him, Rattazzi subjected the government to a dangerous strain; for Garibaldi's popularity was immense, and even those of his countrymen who insisted that no citizen — however distinguished his services — should be permitted to live above the law, and to wage war when he pleased, were as eager as he that Rome should be emancipated.

Untaught by experience, Rattazzi connived at a similar expedition five years later. For several weeks Garibaldi went about openly preaching another crusade. When the French government asked for explanations, Rattazzi had Garibaldi arrested and escorted to Caprera. A dozen men-of-war sailed round and round the rock, forbidding any one to approach or quit it. But one night

Garibaldi escaped in a tiny wherry, and a few days later he led a band of crusaders across the Papal frontier. They met the French troops at Mentana, were worsted and dispersed; and again Garibaldi was locked up in the fortress of Variguano, while one party denounced the government for ingratitude towards the beloved hero, and another denounced it for treating him as a privileged person who might, when the impulse seized him, embroil the country in war. If we regard the acquirement of the methods of constitutional government and of respect for law and order as the chief need of the Italians at that time, we can only regret the agitation and expeditions which Garibaldi conducted, to the detriment of his country's progress.

Meanwhile, in 1866, Venetia had been restored to her kinsfolk, as the result of the brief conflict in which Italy and Prussia allied themselves against Austria. Garibaldi organized another corps of Hunters of the Alps, but the shortness of the campaign prevented him, as in 1859, from going far. In 1870 the war between France and Prussia enabled the Italians to take possession of Rome as soon as the French garrison was withdrawn; so that Italy owed the completion of her unity, not to her own sword, but to a lucky turn in the quarrels of her neighbors.

No sooner had the French Empire collapsed, and the French Republic was seen to be terribly beset by the Germans, than Garibaldi offered his services to her. He was assigned to the command of the Army of the Vosges, a nondescript corps, which more than once gave proof of bravery, although it could not match the superior numbers and discipline of Moltke's men. The French gave him scanty thanks for his services, and at the end of the war he returned home.

During the next ten years he was either at Rome, arraigning the government, the fallen Papacy, and the wastefulness of the monarchy; or he was making triumphal progresses through the land, sure everywhere of being treated as an idol; or he stayed in his Caprera hermitage, inditing letters in behalf of political extremists, Nihilists, fanatics. Yet his popularity did not wane; his countrymen regarded him more than ever as a privileged person, whose senile extravagances were not to be taken too seriously. They loved his intentions; they revered him for the achievements of his prime; and when, on June 2, 1882, he fell asleep in his Caprera home, all Italy put on mourning, and the world, conscious that it had lost a hero, grieved.

On his sixty-fifth birthday (July 4, 1872) he drew his own portrait thus: "A tempestuous life, composed of good and of evil, as I believe of the large part of the world. A consciousness of having sought the good always, for me and for my kind. If I have sometimes done wrong, certainly I did it involuntarily. A hater of tyranny and falsehood, with the profound conviction that in them is the principal origin of the ills and of the corruption of the human race. Hence a republican, this being the system of honest folk, the normal system, willed by the majority, and consequently not imposed with violence and with imposture. Tolerant and not exclusive, incapable of imposing my republicanism by force, on the English, for instance, if they are contented with the government of Queen Victoria. And, however contented they may be, their government should be considered republican. A republican, but evermore persuaded of the necessity of an honest and temporary dictatorship at the head of those nations which, like France, Spain, and Italy, are the victims of a most pernicious Byzantinism. . . . I was copious in praises of the dead, fallen on fields of battle for liberty. I praised less the living, especially my comrades. When I felt myself urged by just rancor against those who wronged me, I strove to placate my resentment before speaking of the offense and of the offender. In every writing of mine, I have always attacked clericalism, more particularly because in it I have always believed that I found the prop of every despotism, of every vice, of every corruption. The priest is the personification of lies, the liar is a thief, the thief is a murderer, — and I could find for the priest a series of infamous corollaries."

Thus he read his own character, and we need not subject it to a searching analysis. In action lay his strength. He trusted instinct against any argument. Hence the single-minded zeal with which he plunged into every enterprise; hence, too, his inability to weigh other policies than his own, and his distrust, often intensified into unreasoning prejudice, of those who differed from him. If his kindly, generous nature often made him the dupe of schemers, the wonder is that they did not beguile him into irreparable excesses. He was saved partly by a thread of common sense and partly by self-respect akin to vanity, which kept him constantly on the alert against being used as a tool. Although modest, he knew so well the grandeur of the part he was playing that he took no pains to dissemble the childlike delight he felt at demonstrations of his popularity. The lifelong champion of democracy, he behaved, in practice, as autocratically as Cromwell; a believer in dictatorships, never able to work successfully as yoke-fellow or subordinate to any one else. Like the dreamers, he could not comprehend that human society, being a growth and not a manufacture, cannot be suddenly lifted by benevolent manifesto or patriotic resolution. He scorned parliamentary debates, he reviled diplomacy, he underrated counsel.

But what he had, he had superlatively: valor, presence of mind, geniality, unselfishness, magnanimity, — he had all these, the qualities of a popular soldier, to a degree which made whoever fought with him worship him. No other man of his time, nor perhaps of any time, inspired so many human beings with personal affection — as distinguished from that devotion which other favorite captains have inspired — as he did. Every one of his soldiers felt that in Garibaldi he had not merely a commander but a brother; every person who approached him acknowledged his fascination.

Strip off Garibaldi's eccentricities, look into his heart, contemplate his achievements, — we behold a hero of the Homeric brood. Again we enter the presence of a man of a few elemental traits, whose habit it was to exhibit his passions without that reserve which belongs to our sophisticated age. Like Achilles, he wept when he was moved, he sulked when he was angry. Equally simple was the mainspring of his action. He obeyed two ideals, and those two of the noblest, — love of liberty and love of his fellow-men: nay, more, he obeyed them as quickly when they led into exile, poverty, and danger as when they led him to a conquest unparalleled in modern history, and to fame in which the wonder and the affection of the world blended in equal parts.

In the making of Italy it was his mission to rouse some of his countrymen to a sense of their patriotic duty, and to lead others to fight for a nation under Victor Emanuel instead of for a faction under Mazzini. Through him, the forces of royalism and of revolution formed an alliance which, although it was almost indispensable to the success of the Italian cause, might never, but for him, have been formed.

Such was Garibaldi, his character, his exploits. Shall we not seek also for the meaning of his career? Shall we not ask, "To what attributes of general human nature had his individuality the key?" That conquest of Sicily was but an episode; long anterior to it was built up the temperament which might have liberated twenty Sicilies, and which found a multitude ready to respond to its least signal.

More than half of our nature is emotion. Men may lie sluggish, they may seem sodden in selfishness, or they may fritter their force away on petty things. But let the hero come, — the Garibaldi, the embodied emotion, — and they will know him as light knows light, or lover his beloved. What just now seemed a dead, sordid mass is tinder, is flame. The craven legions, bewitched and transformed by his example, will follow him anywhere, were it to storm the gates of hell! The immense scope of noble emotion, — is not that the significance, if we seek it, of Garibaldi's marvelous influence? And has it ever been more certainly displayed than in our very century, miscalled prosaic?

PORTRAITS

CARLYLE

Dr. Samuel Johnson, during a long life, cherished an aversion, Platonic rather than militant, for Scotland and the Scotch. Had any one told him that out of the land where oats were fed to men there should issue, soon after his death, a master of romance, an incomparable singer, and a historian without rival, we can well imagine the emphasis with which he would have said, "Tut! tut! sir, that is impossible!" Nevertheless, for the best part of a century Scotland has shed her influence through the world in the genius of Walter Scott, Robert Burns, and Thomas Carlyle; and she has taken sweet vengeance on the burly Doctor himself by creating in James Boswell not only the best of British biographers, but one so far the best that no other can be named worthy to stand second to him. We now celebrate the centenary of the last of these great Scotchmen, — Thomas Carlyle, — and it is fitting that we should survey his life and work.[1]

In a time like our own, when literature on

[1] First printed in *The Forum*, New York, December, 1895.

either side of the Atlantic lacks original energy; when the best minds are busy with criticism rather than with creation; when ephemeral story-tellers and spineless disciples of culture pass for masters, and sincere but uninspired scholars have our respect but move us not, — we shall do well to contemplate anew the man who by his personality and his books has nobly swayed two generations of the English-speaking race, and who, as the years recede, looms more and more certainly as the foremost modern British man of letters. Men may look distorted to their contemporaries, like the figures in a Chinese picture; but Time, the wisest of painters, sets them in their true perspective, gives them their just proportions, and reveals their permanent features in light and shade. And sufficient time has now elapsed for us to perceive that Carlyle belongs to that thrice-winnowed class of literary primates whom posterity crowns. He holds in the nineteenth century a position similar to Johnson's in the eighteenth, and to Milton's in the seventeenth, — each masterful, but in a different way; each typifying his age without losing his individuality; all brothers in preëminence.

When, for convenience' sake, we classify Carlyle among men of letters, we fail to describe him adequately. The phrase suggests too little. Charles Lamb, the lovable, is the true type of men of letters, who illuminate, sweeten, delight, and entertain us. Carlyle was far more: he was a mighty moral force, using many forms of literature — criticism, biography, history, pamphlets — as its organs of expression. He had, as the discerning Goethe said of him, "unborrowed principles of conviction," by which he tested the world. He felt the compulsion of a great message intrusted to him. There rings through most of his utterances the uncompromising "Thus saith the Lord" of the Hebrew prophets, — a tone which, if it do not persuade us, we call arrogant, yet which speaks the voice of conscience to those who give it heed. What, then, was his message? — what those "unborrowed principles of conviction" by which he judged his time?

Born in the poor village of Ecclefechan on December 4, 1795, his childhood and youth were spent amid those stern conditions by which, rather than by affluence, brave, self-reliant, earnest characters are moulded. His parents were Calvinists, to whom religion was the chief concern, and who taught him by example the severe virtues of that grim sect. Next to religion, and its active manifestation in a pious life, they prized education, begrudging themselves no sacrifices by which their son might attend the University of Edinburgh. They wished him to be a minister, but when he came to maturity he recognized his unfitness for that vocation and abandoned it. They acquiesced regretfully, little dreaming that he who refused to be confined in some Annandale pulpit should become the foremost preacher of his age.

Carlyle's reluctance was rooted in conscientious scruples. He began by questioning the authority of his Church; he went on to sift the authority of the Bible. Little by little the whole wondrous fabric of supernatural Christianity crumbled before him. He could not but be honest with himself; he could not but see how Hebrew legend had overgrown the stern ethical code attributed to Moses; how the glosses of Paul and Augustine and a hundred later religionists had changed or perverted the simple teaching of Christ. Awestruck, he beheld the God of his youth vanish out of the world. He wandered in the wilderness of doubt; he wrestled daily and nightly with despair. And then slowly, painfully, after brooding through long years, he saw the outlines of a larger faith emerge from the gloom. He fortified himself by acknowledging that, since righteousness is eternal, it cannot perish when we reject whatever opinions some Council of Westminster, of Trent, or of Nice may have resolved about it.

Only earnest souls who have experienced the wrench which comes when we first break away from the bondage of an artificial religion, and perceive that the moral law may be something very different from dogmas, know the pang it costs. The dread of losing the truth when errors are thrown over — nay, the apparent hopelessness of being able to decide what is truth — causes many to hesitate, and some to turn back. Carlyle was not, of course, the first in Britain to tread the desolate path from Superstition into Rationalism. In the eighteenth century — to go no farther back — two very eminent minds had preceded him; but in both Hume and Gibbon the intellectual predominated over the moral nature, and to temperaments like theirs the pangs of new birth are always less acute. It is because in Carlyle the moral nature preponderated — intense, fiery, and enduring — that he became the spokesman of myriads who since him have had a similar experience.

If we were to hazard a generalization which should sum up the nineteenth century, might we not affirm that the chief business of the century has been to establish a basis of conduct in harmony with what we actually know of the laws governing the universe? Hitherto, for ages together, men have not consciously done this, but they have accepted standards handed down to them by earlier men, who compounded these standards out of little knowledge, much ignorance, legend, and hearsay. Skeptics there have always been, but usually, like the skeptics who flourished in the last century, they have differed from the doubters in ours by the degree of their moral intensity. Whether we turn to Carlyle or to George Eliot, we find each tirelessly busy in substituting for the worn-out tenets of the past, springs of belief and conduct worthy to satisfy a more enlightened conscience.

Here, then, we have the corner-stone of Carlyle's influence. Our world is a moral world; conscience and righteousness are eternal realities, independent of the vicissitudes of any church. If we seek for a definite statement of Carlyle's creed, we shall be disappointed; he never formulated any. After breaking loose from one prison, he would have scoffed at the idea of voluntarily locking himself up in another. He held that to possess a moral sense is to possess its justification; that conscience is a fact transcending logic just as consciousness or life itself does. In the presence of this supreme fact he cared little for its genealogy. The immanence of God was to him an ever-present, awful verity.

Likewise, when we come to examine his philosophy, we discover that he constructed no formal system. He absorbed the doctrine of Kant and his followers, and may be classed, by those who insist that every man shall have a label, among the transcendentalists: but his main interest was the application of moral laws to life, the trial of men and institutions in the court of conscience, rather than the exercise of the intellect in metaphysical speculations. The mystery of evil may not be explained for some ages, if ever; while we argue about it, evil grows: the one indispensable duty for all of us, he would say, is to combat evil in ourselves and in society now and here. The stanch seaman, when his ship founders, does not waste time in meditating why it should be that water will sink a ship, but he lashes together a raft, if haply he may thereby come off safe.

In these respects we behold Carlyle a true representative of his time. Before the vast bulk of sin and sorrow and pain he did not cower; he would fight it manfully. But the smoke of battle darkened him. The spectacle of mankind, dwelling in Eternity, yet ignorant of their heritage, pursuing "desires whose purpose ends in Time;" of souls engaged from dawn to dusk of their swift-fleeting existence, not on soul's business, but on body's business, worshiping idols they know to be false, deceiving, persecuting, slaying each other, — confirmed a tendency to pessimism to which his early Calvinism had predisposed him. But Carlyle's pessimism must not be confounded with Swift's misanthropy, or with Leopardi's blank despair, or with the despicable Schopenhauer's cosmic negation of good. Carlyle was neither cynic nor misanthrope. He might exclaim with Ecclesiastes, " Vanity of vanities, all is vanity!" but he would mean that the ways and works of man are vain in comparison with his possibilities, and with the incalculable worth of righteousness. " Man's unhappiness, as I construe," he says, " comes of his greatness; it is because there is an Infinite in him which, with all his cunning, he cannot quite bury under the Finite. Always there is a black spot in our sunshine: it is even the *Shadow of Ourselves.*"

These being the elements of Carlyle's moral nature, let us look for a moment at the world which he was to test by his " unborrowed principles of conviction." He came on the scene during the decade of reaction which followed the battle of Waterloo. Official Europe, confounding the ambition of Napoleon with the causes underlying the Revolution, supposed that in crushing one it had destroyed the other. The motto of the Old Régime had been *Privilege*, of the New it was *Merit*. The revived political fashions of the eighteenth century, though cut by such elegant tailors as Metternich, Castlereagh, and Polignac, chafed a generation which had grown used to a freer costume. At any time there yawns between the ideals and the practices of society a discrepancy which provokes the censure of the philosopher and the sarcasm of the cynic; but in a time like the Restoration, when some men consciously repudiated and none sincerely believed the system thrust upon them, the chasm between profession and performance must open wider still, revealing not only the noble failures born of earnest but baffled endeavor, but also all the hideous growths of hypocrisy, of deceptions, insincerities, and intellectual fraud. And in very truth the Old Régime resuscitated by Europe's oligarchs was doubly condemned, — first, as being unfitted to the new age; and, secondly, as having marked in the eighteenth century, when it flourished, the logical conclusion of a political and social epoch. In 1820 the trunk and main branches of the tree of Feudalism were dead: he was not a wise man who imagined that the still surviving upper branches would long keep green.

Not alone in the political constitution of society were momentous changes operating. They but represented the attempt of man to work out, in his civic and social relations, ideas which had already penetrated his religion and his philosophy. Distil those ideas to their inmost essence, its name is *Liberty*. The old Church, whether Roman or Protestant, lay rotting at anchor in the land-locked bayou of Authority; and the pioneers of the new convictions, abandoning her and her cargo of antiquated dogmas, had pushed on across intervening morasses to the shore of the illimitable sea; yea, they were launching thereon their skiffs of modern pattern, and resolutely, hopefully steering whither their consciences pointed. Better the storms of the living ocean than the miasma of that stagnant, scum-breeding pool! But a church is of all institutions that to which men cling most stubbornly, paying it lip-service long after its doctrines have ceased to shape their conduct or to lift their aspirations; trying to believe, in spite of their unbelief, that it will continue to be to them a source of strength as it once was to their fathers; preserving forms, but veneering them with contradictory meanings; coming at last to declare that an institution must be kept, if for no other reason than because it once fulfilled the purposes for which it is now inadequate. The aroma of association has for some minds the potency of original inspiration. Who can ponder on life without perceiving that whereas in their business, their possessions, their love, and their hate, men resent dictation; in matters beyond the scope of experience, and consequently beyond proof, — as the conditions of a future life, — men credulously accept the guidance of others quite as ignorant as themselves, from whom in their business or their passions they would submit to no interference?

Needless to say the revived Old Régime intrenched itself behind whatever church it found standing, — in Prussia the Lutheran, in England the Anglican, in Scotland the Calvinist, in the Latin countries the Roman. The ecclesiastical institution might not humanize the masses, but at least it held them in check; it might not spiritualize the classes, but it taught them that in rallying to its support they were best guarding their own privileges. Metternich, whom we call the representative of the Restoration, did not scruple to announce that, as the dangers which threatened Church and State were identical, the Church could be saved only by upholding the State. Not for the first time in history was the priest a policeman in disguise.

Into this world of transition Thomas Carlyle strode with his store of unborrowed principles. Right or wrong, his convictions were his own; therefore they were realities that need not fear a conflict with ghosts of dead convictions and insincerities.

Naturally, one of the first facts that amazed him was the monstrous unreality in that transitional society. By the census the people of Great Britain were rated as Christians; by their acts they seemed little better than barbarians. What availed the Established Church, in which livings were assigned at the pleasure of some dissolute noble, fox-hunting parsons were given the cure of souls, and worldlings or unbelievers rose to be bishops? Could the loudest protestations explain the existence of great, gaunt, brutalized masses, beyond the pale of human charity; every *horse* sleek, well lodged, and well fed, but innumerable *men* dying of hunger or lodged in the almshouse? Can that be true civilization in which the various constituents recognize no interdependence, and only a few usurp benefits which are pernicious unless they be free to all? Respectability, and not virtue, — that, Carlyle declared, was John Bull's ideal, and he opened fire upon its chief allies, Sham and Cant. He spared no prejudices, he respected no institutions. With sarcasm until then unknown in English, he unmasked one artificiality after another, disclosing the cruelty or the hypocrisy which lurked behind it, and setting over against it the true nature of the thing it pretended to be. To interpret such conditions by the criterion of conscience was to condemn them.

But Carlyle's mission was not merely to destroy: he shattered error in order that the clogged fountain of truth might once more gush forth. Before eyes long dimmed with gazing on insincerity, he would hold up shining patterns of sincerity; souls groping for guidance, he would stay and comfort by precedents of strength; hearts pursuing false idols, he would chasten by examples of truth. Men talked — and nowhere more pragmatically than in the churches — as if God, after having imparted his behests to a few Hebrews ages ago, had retired into some remote empyrean, and busied himself no more with the affairs of men. But to Carlyle the immanence of God was an ever-present reality, manifesting itself throughout all history and in every individual conscience, but nowise more clearly than in the careers of great men.

Thus he made it his business to set before his contemporaries models worthy of veneration, for he recognized that worship is a primary moral need. "Great men," he says, "are the inspired (speaking and acting) Texts of that divine *Book of Revelations*, whereof a Chapter is completed from epoch to epoch, and by some named *History*." In this spirit he introduced Goethe, the latest of the heroes, to English readers, as the man who, from amid chaos similar to that which bewildered them, had climbed to a position where life could be lived nobly, rationally, well. "Close your Byron, open your Goethe," was his advice to those in whom Byron's mingled defiance and sentimentality found an echo. He showed in Cromwell how religious zeal is something very different from a phantom faith. He laid bare the truth in Mahomet. He made Luther live again. And all to the end that he might convince his dazed contemporaries that in no age, if we look deeply, shall we look in vain for concrete, living examples of those qualities which are indispensable to right action; that salvation — the purging of the character — is won by exercising virtues, and not by conforming to a stereotyped routine; that the authority of conscience is a present fact, not a mere mechanism which God wound up and gave to the Hebrews, and has been transmitted in poor repair by them to us. As an antidote to sterilizing doubt, Carlyle prescribed the simple remedy which sums up the wisdom of all the sages: "Do the Duty which lies nearest thee, which thou knowest to be a Duty! Thy second Duty will already have become clearer." In this fashion did Carlyle discharge his mission as a moral regenerator. We live as individuals, and to the individual conscience he made his appeal, caring little for the organization of principles into institutions. Rather, like every individualist, did he incline to deprecate the numbing effect of institutions. Let each unit be righteous, in order that whatever the collective units shall establish may be righteous too.

Bearing this in mind, we shall understand Carlyle's attitude toward the great social and intellectual movements of his time. The watchword which had inspired generous minds at the end of the last century was *Liberty*, and after the thunders of the Napoleonic wars that had drowned it died away, it rang out its summons more clearly than before, never again to be quite deadened, despite all the efforts of the Old Régime. The application of the theory of Liberty to government resulted in setting up Democracy as the ideal political system. Since every citizen in the State bears, directly or indirectly, his fraction of the burden of taxation, and since he is affected by the laws, and interested, even to the point of laying down his life, in the preservation of his country, Democracy declares that he should have an equal part with every other citizen in determining what the taxes and policy of his State shall be; and it thrusts upon him the responsibility of choosing his own governors and representatives. To Carlyle this ideal seemed a chimera. Honest, just, and intelligent government is of all social contrivances the most difficult: by what miracle, therefore, shall the sum of the opinions of a million voters, severally ignorant, be intelligent? As well blow a million soap-bubbles, each thinner than gossamer, and expect that collectively they will be hard as steel! Or, admitting that the representatives

Demos chooses be not so incompetent as itself, how shall they be kept disinterested? Their very numbers not only make them unmanageable, but so divide responsibility that any individual among them can shift from his own shoulders the blame for corrupt or harmful laws. Moreover, popular government means party government, and that means compromise. To Carlyle, principles were either right or wrong, and between right and wrong he saw no neutral ground for compromise. Party government cleaves to expediency, which at best is only a half-truth; but half-truth is also half-error, and any infinitesimal taint of error vitiates the truth to which it clings. Finally, Democracy substitutes a new, many-headed tyranny — more difficult to destroy because many-headed — for the tyranny it would abolish.

Such objections Carlyle urged with consummate vigor. He foresaw, too, many of the other evils which have accompanied the development of this system to impair its efficacy, such as the rise of a class of professional politicians, of political sophists, of corrupt "bosses," expert in the art of wheedling the ignorant many, and thereby of frustrating the initial purpose of the system. His opposition did not spring from desire to see the masses down-trodden, but from conviction that they need guidance and enlightenment, and that they are therefore no more competent to choose their own law-makers than children are to choose their own teachers. In knowledge of public affairs Demos is still a child, innocent, well-intentioned, if you will; but ignorant, and by this system left to the mercy of the unscrupulous.

This brings us to consider the charge that Carlyle, in his exaltation of the Strong Man, worshiped crude force. Let us grant that on the surface the accusation seems plausible; but when we seek deeper, we shall discover that he exalts Cromwell and Frederick, not because they were despots, but because, in his judgment, they knew better than any other man, or group of men, in their respective countries, how to govern. Their ability was their justification; their force, but the symbol of their ability. "Weakness" — Carlyle was fond of quoting — "is the only misery." What is ignorance but weakness (through lack of training) of the intellect? In the incessant battle of life, — and few men have been more constantly impressed than Carlyle by the battle-aspect of life, — weakness of whatever kind succumbs to strength. Evil perpetually marshals its forces against Good, — positive, aggressive forces, to be overcome neither by inertia, nor indifference, nor half-hearted compromise, but by hurling stronger forces of Good against them. Interpreting Carlyle's views thus, we perceive why he extolled the Strong Man and distrusted the aggregate ignorance of Democracy. Furthermore, we must not forget that he never considered politics the prime business of life : first, make the masses righteous, next, enlightened, and then they will naturally organize a righteous and enlightened government. When Carlyle rejoined to the zealots of Democracy or other panaceas, "Adopt your new system if you must, will not the same old human units operate it? Were it not wiser to perfect them first?" — he antagonized the spirit of the age : wisely or not, only time can show. Those of us who would reject his arguments would nevertheless admit that Democracy is still on trial.

With equal fearlessness he attacked the cheap optimism based on material prosperity, which brags of the enormous commercial expansion made possible by the invention of machinery; which boasts of the rapid increase in population — so many more million mouths to feed and bodies to clothe, and so much more food and raiment produced — from decade to decade. These facts, he insisted, are not of themselves evidences of progress. Your inventions procure greater comfort, a more exuberant luxury; but do comfort and luxury necessarily build up character? — do they not rather unbuild it? Are your newly bred millions of bodies more than bodies? Take a census of souls, has *their* number increased? Though your steam-horse carries you fifty miles an hour, have you thereby become more virtuous? Though the lightning bears your messages, have you gained bravery? Of old, your aristocracy were soldiers: is the brewer who rises from his vats to the House of Lords — is any other man owing his promotion to the tradesman's skill in heaping wealth — more worshipful than they? Let us not say that this amazing industrial expansion may not conduce to the uplifting of character; but let us strenuously affirm that it is of itself no indication of moral progress, and that, if it fail to be accompanied by a corresponding spiritual growth, it will surely lead society by the Byzantine high-road to effeminacy, exhaustion, and death.

A different gospel, this, from that which Carlyle's great rival, Macaulay, was preaching, — Macaulay, who lauded the inventor of a useful machine above all philosophers! Different from the optimism — which gauges by bulk — of the newspapers and the political haranguers! Different, because true! Yet, though it sounded harsh, it stirred consciences, — which smug flatterings and gratulations can never do; and it gave a tremendous impetus to that movement which has come to overshadow all others, — the movement to reconstruct society on a basis, not of privilege, not of bare legality, but of mutual obligations.

Any inventory, however brief, of Carlyle's substance, would be incomplete without some reference to his quarrel with Science. To Science a large part of the best intelligence of our age has been devoted, — a sign of the breaking away of the best minds from the cretinizing quibbles of theology into fields where knowledge can be ascertained. It is a truism that Science has advanced farther in our century than in all preceding time. By what paradox, then, should Carlyle slight its splendid achievements? Was it not because he revolted from the materialistic tendency which he believed to be inseparable from Science, a tendency which predominated a generation ago more than it does to-day? Materialism Carlyle regarded as a Gorgon's head, the sight of which would inevitably petrify man's moral nature.

Moreover, Carlyle's method differed radically from that of the scientific man, who describes processes and investigates relations, but does not explain causes. Pledged to his allegiance to tangible facts, the man of science looks at things serially, pays heed to an individual as a link in an endless chain rather than as an individual, lays emphasis on averages rather than on particulars. To him this method is alone honest, and, thanks to it, a single science to-day commands more authenticated facts than all the sciences had fifty years ago. But there are facts of supreme importance which, up to the present at least, this method does not solve. The mystery of the origin of life still confronts us. Consciousness, the Sphinx, still mutely challenges the caravans which file before her. The revelations of Science seem, under one aspect, but descriptions of the habitations of life from the protoplasmic cell up to the human body. Immense though the value of such a register be, we are not deceived into imagining that it explains ultimates. How came life into protoplasm at all? Whence each infinitesimal increment of life, recognizable at last in the budding of some new organ? And when we arrive at man, whence came his personality? Each of us is not only one in a genealogical series stretching back to the unreasoning, conscienceless *amœba*, but a clearly defined individual, a little world in himself, to whom his love, his sorrow, his pain and joy and terror, transcend in vividness all the experiences of all previous men: a microcosm, having its own immediate relations — absolute relations — with the infinite macrocosm. Science, bent on establishing present laws, measures by æons, counts by millions, and has warrant for ignoring your brief span or mine; but to you and me these few decades are all in all. However it may fare with the millions, you and I have vital, pressing needs, to supply which the experience of the entire animal kingdom can give us no help. Upon these most human needs Carlyle fastened, to the exclusion of what he held to be unnecessary to the furtherance of our spiritual welfare. He busied himself with ultimates and the Absolute. Not the stages of development, but the development attained; not the pedigree of conscience, but conscience as the supreme present reality; not the species, but the individual, — were his absorbing interests.

Thus we see how Carlyle approached the great questions of life invariably as a moralist. Mere erudition, which too often tends away from the human, did not attract him. Science, which he beheld still unspiritualized, he undervalued: what boots it to know the "mileage and tonnage" of the universe, when our foremost need is to build up character? In politics, in philosophy, in religion, likewise, he set this consideration above all others: before its august presence outward reforms dwindled into insignificance.

Such was the substance of Carlyle's message. Remarkable as is its range, profound as is its import, it required for its consummation the unique powers of utterance which Carlyle possessed. Among the masters of British prose he holds a position similar to that of Michael Angelo among the masters of painting. Power, elemental, titanic, rushing forth from an inexhaustible moral nature, yet guided by art, is the quality in both which first startles our wonder. The great passages in Carlyle's works, like the Prophets and Sibyls of the Sixtine Chapel, have no peers: they form a new species, of which they are the only examples. They seem to defy the ordinary canons of criticism; but if they break the rules it is because whoever made the rules did not foresee the possibility of such works. Transcendent Power, let it take whatever shape it will, — volcano, torrent, Cæsar, Buonarotti, Carlyle, — proclaims: "Here I am, — a fact: make of me what you can! You shall not ignore me!"

Of Carlyle's style we may say that, whether one likes it or not, one can as little ignore it as fail to perceive that he makes it serve, with equal success, whatever purpose he requires. It can explain, it can laugh, it can draw tears; it can inveigh, argue, exhort; it can tell a story or preach a sermon. Carlyle has, it is computed, the largest vocabulary in English prose. His endowment of imagination and of humor beggars all his competitors. None of them has invented so many new images, or given to old images such fresh pertinence. Your first impression, on turning to other writings after his, is that they are pale, and dim, and cold: such is the fascination inalienable from power. Excess there may be in so vehement a genius; repetition there must be in utterances poured out during sixty years; an individuality so intense must have an equally individual manner; but there is, rightly speaking, no mannerism, for mannerism implies affectation, and Carlyle's primal instinct was sincerity. His expression is an organic part of himself, and shares his merits and defects.

Carlyle won his first reputation as a historian; singularly enough, his achievements in history have temporarily suffered a partial eclipse. Teachers in our colleges refer to them dubiously or not at all. Does the fault lie with these same teachers, or with Carlyle? A glance at the methods of the school of historical students which has sprung up during the last generation will explain the disagreement.

History, like every other branch of intellectual activity, has responded to the doctrines of Evolution. That most fertile working hypothesis has proved, when applied here, not less fruitful than in other fields. It has caused the annals of the past to be reinvestigated, every document, record, and monument to be gathered up, and the results have been set forth from the new point of view. Evolutionary science, as we saw above, fixes its attention primarily on the processes of development, and regards the individual, in comparison with a species or the race, as a negligible quantity. A similar spirit has guided historical students. They have turned away from "great captains with their guns and drums," away from figure-head monarchs, away from the achievements of even the mightiest individuals, to scrutinize human action in its collective forms, the rise and supremacy and fall of institutions, the growth of parties, the waxing and waning of organisms like Church or State, in whose many-centuried existence individual careers are swallowed up. Using the methods of Science, these students have persuaded themselves that history also is a science, which, in truth, it can never be. Judicial temper, patience, veracity, — the qualities which they rightly magnify, — were not invented by them, nor are these the only qualities required in writing history. Speaking broadly, facts lie within the reach of any diligent searcher. But a fact is a mere pebble in a brook until some David comes to put it in his sling. True history is the arrangement and interpretation of facts, and — more difficult still — insight into motives : for this there must be art, there must be imagination.

To the disciples of the "scientific school" it may be said that the heaping up of great stores of facts — the collection of manuscripts, the cataloguing of documents, the shoveling all together in thick volumes prefaced by forty pages of bibliography, each paragraph floating on a deep, viscous stream of notes, each volume bulging with a score of appendices — is in no high sense history, but the accumulation of material therefor. It bears the same relation to history as the work of the quarryman to that of the architect; most worthy in itself, and evidently indispensable, but not the same. Stand before some noble edifice, — Lincoln Cathedral, for instance, with its incomparable site, its symmetry and majestic proportions: scan it until you feel its personality and realize that this is a living idea, the embodiment of strength and beauty and aspiration and awe, — and you will not confound the agency of the stone-cutters who quarried the blocks with that of the architect in whose imagination the design first rose. Neither should there be confusion between the historical hodman and the historian.

Indubitably, history of the highest kind may be written from the evolutionist's standpoint, but as yet works of the lower variety predominate. Therefore, in a time when the development of institutions chiefly commands attention, Carlyle, who magnifies individuals, will naturally be neglected. But in reality, histories of both kinds are needed to supplement each other. All institutions originate and exist in the activities of individuals. The hero, the great man, makes concrete and human what would otherwise be abstract. Environment does not wholly explain him. It is easy to show wherein he resembles his fellows; that difference from them which constitutes his peculiar, original gift is the real mystery, which the study of resemblances cannot solve. Men will cease to be men when personality shall lose its power over them.

Accepting, therefore, the inherent antagonism in the two points of view, — antagonism which implies parity and not the necessary extinction of one by the other, — we can judge Carlyle fairly. Among historians he excels in vividness. Perhaps more than any other who has attempted to chronicle the past, he has visualized the past. The men he describes are not lay figures, with wooden frames and sawdust vitals, to be called Frenchmen or Germans or Englishmen according as a different costume is draped upon them; but human beings, each swayed by his own passions, striving and sinning, and incessantly alive. They are actors in a real drama: such as they are, Carlyle has seen them; such as he has seen, he depicts them. To go back to Carlyle from one of the "scientific historians" is like passing from a museum of mummies out into the throng of living men. If his portraits differ from those of another artist, it does not follow that they are false. In ordinary affairs, two witnesses may give a different report of the same event, yet each may, from his angle of observation, have given exact testimony. Absolute truth, who shall utter it? Since history of the highest, architectonic kind is interpretation, its value must depend on the character of the interpreter. Not to be greatly esteemed, we suspect, are those grubbers among the rubbish heaps who imagine that Carlyle's interpretation of the French Revolution, or of Cromwell, or of Frederick, may be ignored. Character, insight, and imagination went to the production of works like these: they require kindred gifts to be appreciated.

Neither of Carlyle's portrait gallery, unparalleled in range, in which from each picture an authentic human face looks out at us; nor of his masterpieces of narration, long since laureled even by the unwilling, — is there space here to speak. In portraiture he used Rembrandt's methods: seizing on structural and characteristic traits, he displays them in strong, full light, and heightens the effect by surrounding them with shadows. As a biographer he succeeded equally well in telling the story of Schiller and that of John Sterling: the latter a most difficult task, as it must always be to make intelligible to strangers a beautiful character whose charm and force are felt by his friends, but have no proportionate expression in his writings. As an essayist he has left models in many branches: "Mirabeau," "Johnson," "Goethe," "Characteristics," "Burns," "History," stand as foothills before his more massive works. His is creative criticism, never restricted, like the criticism of the schools, to purely literary, academic considerations, but penetrating to the inmost heart of a book or a man, to discover what deepest human significance may there be found. A later generation has, as we have noted, adopted a different treatment in all these fields: bending itself to trace the ancestry and to map out the environment of men of genius; concentrating attention on the chain rather than its links; necessarily belittling the individual to aggrandize the mass. It behooves us, while we recognize the value of this treatment as a new means to truth, not to forget that it is not the only one. By and by — perhaps the time is already at hand — we shall recognize that the other method, which deals with the individual as an ultimate rather than in relation to a series, which is human rather than abstract, cannot be neglected without injury to truth. Either alone is partial; each corrects and enlightens the other.

Meanwhile we will indulge in no vain prophecies as to Carlyle's probable rank with posterity. That a man's influence shall be permanent depends first on his having grasped elemental facts in human nature, and next on his having given them an enduring form. Systems struggle into existence, mature, and pass away, but the needs of the individual remain. Though we were to wake up to-morrow in Utopia, the next day Utopia would have vanished, unless we ourselves had been miraculously transformed. To teach the individual soul the way of purification; to make it a worthy citizen of Eternity which laps it around; to kindle its conscience; to fortify it with courage; to humanize it with sympathy; to make it true, — this has been Carlyle's mission, performed with all the vigor of a spirit " in earnest with the universe," and with intellectual gifts most various, most powerful, most rare. It will be strange if, in time to come, souls with these needs, which are perpetual, lose contact with him. But, whatever befall in the future, Carlyle's past is secure. He has influenced the *élite* of two generations: men as different as Tyndall and Ruskin, as Mill and Tennyson, as Browning and Arnold and Meredith, have felt the infusion of his moral force. And to the new generation we would say: " Open your ' Sartor ; ' there you shall hear the deepest utterances of Britain in our century on matters which concern you most; there, peradventure, you shall discover yourselves."

TINTORET.[1]

I. HIS LIFE.

WE have no authentic biography of Tintoret. The men of his epoch hungered for fame, but it was by the splendor of their genius, and not by the details of their personal lives, that they hoped to be known to posterity. The days of judicious Boswells and injudicious Froudes had not then come to pass; so that we are now as ignorant of the lives of the painters of the great school which flourished at Venice during the sixteenth century as of the lives of that group of poets who flourished in England during the reigns of Elizabeth and James I. Nevertheless, Providence sees to it that nothing essential be lost; and, in the absence of memoirs, the masterpiece itself becomes a memoir for those who have insight. In art, works which proceed from the soul, and not from the skill, are truthful witnesses to the character of the artist. "For by the greatness and beauty of the creatures proportionably the maker of them is seen." It is not wholly to be regretted, therefore,

[1] First printed in *The Atlantic Monthly*, June, 1891.

that the meagreness of our information concerning Tintoret compels us to study his paintings the more earnestly. The lives of artists are generally scanty in those adventures and dramatic incidents which make entertaining biographies. Men of action express their character in deeds: poems, statues, paintings, are the deeds of artists. Blot out a few pages of history, and what remains of Hannibal or Scipio? But we should know much about Michael Angelo or Raphael from their paintings, had no written word about either come down to us.

The year of Tintoret's birth is variously stated as 1512 and 1518. Even his name has been a cause of dispute to antiquaries; but since he was content to call and sign himself Jacopo (or Giacomo) Robusti, we may accept this as correct. His father was a dyer of silk (*tintore*), and as the boy early helped at that trade he got the nickname *il tintoretto*, "the little dyer." Vasari, also born in 1512, is the only contemporary who furnishes an account of Tintoret. Unsatisfactory and well-nigh ridiculous it is, if we remember that by 1574, when Vasari died, Tintoret had already produced many of his masterpieces. Yet the Florentine painter-historian did not accord to him so much as a separate chapter in his "Lives of the Most Excellent Painters," but inserted his few pages of criticism and gossip, as if by an afterthought, in the sketch of the forgotten Battista Franco. Since much that has been subsequently written about Tintoret is merely a repetition of Vasari's shallow opinions, which created a mythical Tintoret, just as English reviewers created a mythical "Johnny Keats," long believed to be the real Keats, I quote a few sentences from Vasari.

"There still lives in Venice," he says, "a painter called Jacopo Tintoretto, who has amused himself with all accomplishments, and particularly with playing music and several instruments, and is, besides, pleasing in all his actions; but in matters of painting he is extravagant, full of caprice, dashing, and resolute, the most terrible brain that painting ever had, as you may see in all his works, and in his compositions of fantastic subjects, done by him diversely and contrary to the custom of other painters. Nay, he has capped extravagance with the novel and whimsical inventions and odd devices of his intellect, which he has used haphazard and without design, as if to show that this art is a trifle. . . . And because in his youth he showed himself in many fair works of great judgment, if he had recognized the great endowment which he received from nature, and had fortified it with study and judgment, as those have done who have followed the fine manner of his elders, and if he had not (as he has done) cut loose from practiced rules, he would have been one of the greatest painters that ever Venice had; yet, for all this, we would not deny that he is a proud and good painter, with an alert, capricious, and refined spirit." [1]

Evidently, the originality of this "terrible" Tintoret could not be understood by Vasari, who imagined that he followed successfully the fine manner of his elders in the academic proprieties. But there is no hint that Tintoret heeded this generous advice. Perhaps it came too late, — at threescore years one's character and methods are no longer plastic; perhaps it had been too often reiterated, for Tintoret had been assured from his youth up that, if he would only be instructed by his fellow-artists, he might hope to become a great painter like them. But, from the first glimpse we get of this perverse Tintoret to the last, one characteristic dominates all, — obedience to his own genius. Censure, coaxing, fashion, envy, popu-

---

[1] Vasari's condescending estimate of Tintoret may remind some readers of Voltaire's patronizing estimate of Shakespeare: "It seems as though nature had mingled in the brain of Shakespeare the greatest conceivable strength and grandeur with whatever witless vulgarity can devise that is lowest and most detestable;" and much more of the same kind about the "intoxicated barbarian," which will seem pitiful or amusing according to the humor of the reader.

larity, seem never to have swerved him. Like every consummate genius, he drew his inspiration directly from within. "Conform! conform! or be written down a fool!" has always been the greeting of the world to the self-centred, spirit-guided few. "Right or wrong, I cannot otherwise," has been their invariable reply.

By the time that Tintoret made his first essays in painting, the Venetian school was the foremost in the world. The great Leonardo had died in France, leaving behind him in Lombardy a company of pupils who were rapidly enslaved by a graceless mannerism. Even earlier than this, the best talents of Umbria had wandered into feeble eccentricities, or had been absorbed by Raphael's large humanism. Raphael himself was dead, at the height of his popularity and in the prime of his powers, and his disciples were hurrying along the road of imitation into the desert of formalism. Michael Angelo alone survived in central Italy, a Titan too colossal, too individual, to be a schoolmaster, although there were many of the younger brood (Vasari among them) who called him *Maestro*, and fancied that the grimaces and contortions they drew sprang from force and grandeur such as his. But in Venice painting was flourishing; there it had the exuberance and the strength, the joyousness and the splendor, of an art approaching its meridian. John Bellini, the eldest of the great Venetians, had died; but not before there had issued from his studio a wonderful band of disciples, some of whom were destined to surpass him. Giorgione, one of these, had been cut off in his thirty-fourth year, having barely had time to give to the world a few handsels of his genius. The fame of Titian had risen to that height where it has ever since held its station. A troop of lesser men — lesser in comparison with him — were embellishing Venice, or carrying the magic of her art to other parts of Italy.

The tradition runs that the boy Tintoret amused himself by drawing charcoal figures on the wall, then coloring them with his father's dyes: whence his parents were persuaded that he was born to be a painter. Accordingly, his father got permission for him to work in Titian's studio, the privilege most coveted by every apprentice of the time. His stay there was brief, however; hardly above ten days, if the legend be true which tells how Titian returned one day and saw some strange sketches, and how, learning that Tintoret had made them, he bade another pupil send him away. Some say that Titian already foresaw a rival in the youthful draughtsman; others, that the figures were in a style so contrary to the master's that he discerned no good in them, and judged that it would be useless for Tintoret to pursue an art in which he could never excel. Let the dyer's son go back to his vats: there he could at least earn a livelihood. We are loth to believe that Titian, whose reputation was established, could have been moved by jealousy of a mere novice: we must remember, nevertheless, that even when Tintoret had come to maturity, and was reckoned among the leading painters of Venice, Titian treated him coldly, and apparently thwarted and disparaged him. Few artists, indeed, have risen quite above the marsh-mists of jealousy. Their ambition regards fame as a fixed quantity, and, like Goldsmith, they look upon any one who acquires a part of this treasure as having diminished the amount they can appropriate for themselves. But in Tintoret's great soul envy could find no place. "Enmities he has none," as Emerson says of Goethe. "Enemy of him you may be: if so, you shall teach him aught which your good-will cannot, were it only what experience will accrue from your ruin. Enemy and welcome, but enemy on high terms. He cannot hate anybody; his time is worth too much."

Under whom Tintoret studied, after being thrust off by Titian, we are not told. Probably he had no acknowledged preceptor except himself. Already his aim was at the highest. On the wall of his studio he blazoned the motto, "*The drawing of Michael Angelo and the coloring of Titian."* To blend the excellence of each in a supreme unity, — that was his ambition. Titian might shut him out from personal instruction, but Titian's works in the churches and palaces were within reach. Tintoret studied them, copied them, and conjured from them the secret their master wished to hide. Having procured casts of Michael Angelo's statues in the Medicean Chapel at Florence, he made drawings of them in every position. Far into the night he worked by lamplight, watching the play of light and shade, the outlines and the relief. He drew also from living models, and learned anatomy by dissecting corpses. He invented " little figures of wax and of clay, clothing them with bits of cloth, examining accurately, by the folds of the dresses, the position of the limbs; and these models he distributed among little houses and perspectives composed of planks and cardboard, and he put lights in the windows." From the rafters he suspended other manikins, and thereby learned the foreshortening proper to figures painted on ceilings and on high places. So indefatigable, so careful, was this man, who is known to posterity as "the thunderbolt of painters"! In his prime, he astonished all by his power of elaborating his ideas at a speed at which few masters can even sketch; but that power was nourished by his infinite painstaking in those years of obscurity.

Wherever Tintoret might learn, thither he went. Now we hear of him working with the masons at Cittadella; now taking his seat on the bench of the journeymen painters in St. Mark's Place; now watching some illustrious master decorating the façade of a palace. No commission was too humble for him: who knows how many signboards he may have furnished in his 'prentice days? His first recorded works were two portraits, — of himself holding a bas-relief in his hand, and of his brother playing a cithern. As the custom then was, he exhibited these in the Merceria, that narrow lane of shops which leads from St. Mark's to the Rialto. What the latest novel or yesterday's political speech is to us, that was a new picture to the Venetians. Their innate sense of color and beauty and their familiarity with the best works of art made them ready critics. They knew whether the colors on a canvas were in harmony, as the average Italian of to-day can tell whether a singer keeps the key, and doubtless they were vivacious in their discussions. Tintoret's portraits attracted attention. They were painted with nocturnal lights and shadows, "in so terrible a manner that they amazed every one," even to the degree of suggesting to one beholder the following epigram:

> "Si Tinctorectus noctis sic lucet in umbris,
> Exorto faciet quid radiante die?"[1]

Soon after, he displayed on the Rialto bridge another picture, by which the surprise already excited was increased, and he began thenceforward to get employment in the smaller churches and convents. Important commissions which brought wealth and honors were reserved for Titian and a few favorites; but Tintoret rejected no offer. Only let him express those ideas swarming in his imagination: he asked no further recompense. He seems to have been early noted for the practice of taking no pay at all, or only enough to provide his paints and canvas, — a practice which brought upon him the abuse of his fellows, who cried out that he would ruin their profession. But there was then no law to prohibit artist or artisan from working for any price he chose, and Tintoret, as usual, took his own course.

At last a great opportunity offered. On each side of the high altar of the Church of Sta. Maria dell' Orto was a bare space, nearly fifty feet high and fifteen or twenty feet broad. "Let me paint you two pictures," said Tintoret to the friars, who laughed at the extravagant proposal. "A whole year's income would not suffice for such an under-

---

[1] If Tintoret shines thus in the shades of night, what will he do when radiant day has risen?

taking," they replied. "You shall have no expense but for the canvas and colors," said Tintoret. "I shall charge nothing for my work." And on these terms he executed "The Last Judgment" and "The Worship of the Golden Calf." The creator of those masterpieces could no longer be ignored. Here was a power, a variety, which hostility and envy could not gainsay: they must note, though they refused to admire. It was in 1546, or thereabouts, that Tintoret uttered this challenge. In a little while he had orders for four pictures for the School of St. Mark; one of which, "St. Mark Freeing a Fugitive Slave," soon became popular, and has continued so. "Here is coloring as rich as Titian's, and energy as daring as Michael Angelo's!" visitors still exclaim. Other commissions followed, until there came that which the Venetian prized above all others, — an order to paint for the Ducal Palace.

As the patriotic Briton aspires to a monument in Westminster Abbey, as the Florentine covets a memorial in Santa Croce, so the Venetian artist coveted for his works a place in the Palace of the Doges. That was his Temple of Fame. His dream, however, soared beyond the gratification of personal ambition: he desired that through him the glory and beauty of Venice might be enhanced and immortalized. This devotion to the ideal of a city, this true patriotism, has unfortunately almost disappeared from the earth. The very conception of it is now unintelligible to most persons. The city where you live — New York, Boston, London — you value in proportion as it affords advantages for your business, objects for your comfort and amusement; but you quit it without compunction if taxes be lower and trade brisker elsewhere. You are interested in its affairs just in so far as they affect your own. When you build a dwelling or a factory, you do not inquire whether it will improve or injure your neighbor's property, much less whether it will be an ornament to the city; you need not even abate a nuisance until compelled to do so by the law.

But to the noble-minded Venetian, his city was not merely a convenience: it was a personality. Venezia was a spiritual patroness, a goddess who presided over the destiny of the State; he and every one of his fellow-citizens shared the honor and blessing of her protection. She had crowned with prosperity the energy and piety, the rectitude and justice, of his ancestors through many centuries. Every act of his had more than a personal, more even than a human, bearing. *How would it affect her?* — that was his test. He could do nothing unto himself alone; for good or for ill, what he did reacted upon the community, upon the ideal Venezia. The outward city — the churches, palaces, and dwellings — was but the garment and visible expression of that ideal city. Venezia had blessed him, and he was grateful; she was beautiful, and he loved her. His gratitude impelled him to deeds worthy of her protection; his love blossomed in gifts that should increase her beauty.

This reverence and devotion have, as I remarked, vanished from among men; yet in this ideal beams the conception of the true commonwealth. Observe that those three cities which held such an ideal before them have bequeathed to us the most precious works of beauty. Athens, Florence, Venice, — these are the Graces among the cities. At Karnak, at Constantinople, at Rome, at Paris, you will behold stupendous ruins or imposing monuments commemorating the pride and power of individual Pharaohs, Sultans, Cæsars, Popes, and Napoleons, but you will not find the spirit which was worshiped by the beautifying of the Acropolis, and of republican Florence, and of Venice. In which modern city will the most diligent search discover it?

Tintoret, then, had at last earned the privilege of consecrating his genius to Venezia. His first work for her seems to have been a portrait of the reigning Doge.[1] Then he painted two historical

[1] It is interesting to know that the price regularly paid to

subjects. — "Frederick Barbarossa being crowned by Pope Adrian," and "Pope Alexander III excommunicating Frederick Barbarossa;" and "The Last Judgment," destroyed by the fire of 1557. Not long thereafter began his employment by the brothers of the confraternity of San Rocco. For their church, about 1560, he painted two scenes in the life of St. Roch, and then he joined in competition for a ceiling painting for the Salla dell' Albergo in the School itself. The brothers called for designs, and upon the appointed day Paul Veronese, Andrea Schiavone, Giuseppe Salviati, and Federigo Zuccaro submitted theirs. But Tintoret had outsped them, and when his design was asked for he caused a screen to be removed from the ceiling, and lo! there was a finished picture of the specified subject. Brothers and competitors were astonished, and not greatly pleased. " We asked for sketches," said the former. " That is the way I make my sketches," replied Tintoret. They demurred; but Tintoret presented the picture to the School, one of whose rules made it obligatory that all gifts should be accepted. The displeasure of the confraternity soon passed away, and Tintoret

Titian and Tintoret for state portraits was twenty-five ducats (about thirty-one dollars). Painters who have not a hundredth part of the genius of either Titian or Tintoret now receive one hundred times that sum.

was commissioned to furnish whatever paintings should be required in future. An annual salary of one hundred ducats was bestowed upon him, in return for which he was to give at least one painting a year. Generously did he fulfil the contract; for at his death the School possessed more than sixty of his works, for which he had been paid but twenty-four hundred and forty-seven ducats.

In 1577 a fire in the Ducal Palace destroyed many of the paintings, and when the edifice was restored the government looked for artists to replace them. Titian being dead, his opposition had no longer to be overcome; yet even now Tintoret had to compete with men of inferior powers, but of stronger influence. Nevertheless, to him and Paul Veronese was assigned the lion's share of the undertaking, and for ten years those two men labored side by side, in noble rivalry, to eternize the beauty and the glory of Venice. In 1588, owing to the death of Paul Veronese, who with Francesco Bassano had been commissioned to paint a "Paradise" in the Hall of the Grand Council, the work was transferred to Tintoret, who devoted to it the last six years of his life, and left in it the highest expression not only of his genius, but of Italian painting.[1] Old age robbed him of none

[1] Has any one remarked that when Tintoret was painting the

of his energy, but added sublimity to his imagination, and interfused serenity and mellowness throughout his work. Still teeming with plans, he died of a gastric trouble, after a fortnight's illness, on the 31st of May, 1594.[1]

With this clue, spun from the discursive records of Ridolfi (whose *Meraviglie dell' Arte* was first published in 1648), we can pass through the labyrinth of Tintoret's career. There are, besides, several anecdotes which help us to know the man's personality better: if all be not authentic, at least all agree in attributing to him certain well-defined traits.

As a workman, as we have seen, Tintoret was indefatigable. His lifelong yearning was not for praise, but for opportunity to work. Modesty he had to a degree unrecorded of any other painter, although none seems to have been more confident of his own powers.[2] Like Shakespeare, he wrought

"Paradise," Cervantes, Spain's spokesman before the nations, Montaigne, the largest figure in French literature, and Shakespeare, paragon not of England only, but of the world, were his contemporaries? Those four might have met in his studio; and Science might have furnished three peerless representatives, — Bacon, Galileo, and Kepler.

[1] Tintoret is buried in the church of Santa Maria dell' Orto.

[2] Two instances are worthy of record. Having agreed to paint a large historical picture for the Doges' Palace, he said to the procurators, "If any other shall, within the space of two years, paint a better picture of this subject, you shall take his and reject

his masterpieces swiftly, and left them to their fate, because his imagination, like Shakespeare's, was already on the wing for higher quarry. There was in the man an inflexible dignity, born of self-respect, which neither the allurements of popularity nor the flattery of the great could bend. When invited by the Duke of Mantua to go to that city and execute some paintings, Tintoret replied that wherever he went his wife wished to accompany him; at which the Duke bade him bring his wife and family, had them conveyed to Mantua in a state barge, and entertained them at his palace " at magnificent expense for many days." He urged Tintoret to settle there; but the Venetian could not be persuaded to renounce his allegiance to Venice. He saw that titles would add nothing to his fame, and refused an offer of knighthood from Henry III of France. Princes and grandees and illustrious visitors to Venice went to his house; but though he received them courteously, he sought no intimacy with them. His time was too precious, his projects were too earnest, to allow of aristocratic dissipation. He had a keen sense

---

mine." At first his enemies spoke so censuringly of his " St. Mark Freeing the Fugitive Slave " that the brethren hesitated whether to accept it; whereupon Tintoret had it brought back to his studio. Afterwards the brethren repented, begged for its return, and ordered three other pictures.

of humor, which displayed itself now in some ready reply, now in genial conversation with his familiars. Ridolfi relates that certain prelates and senators who visited him whilst he was making sketches for the "Paradise" asked him why he worked so hurriedly, whereas John Bellini and Titian had been deliberate and painstaking. "The old masters," said Tintoret, "had not so many to bother them as I have." At another time, at a gathering of amateurs, a woman's portrait by Titian was lauded. "That's the way to paint," said one of the critics. Tintoret went home, took a sketch by Titian and covered it with lampblack, painted a head in Titian's manner on the same canvas, and showed it at the next meeting of these amateurs. "Ah, there's a real Titian!" they all agreed. Tintoret rubbed off the lampblack from the original sketch and said: "This, gentlemen, is indeed by Titian; that which you have admired is mine. You see now how authority and opinion prevail in criticism, and how few there are who really understand painting."

Pietro Aretino, that depraved adventurer and most successful blackmailer in literature, was one of Titian's intimates and partisans. He wished, nevertheless, to have his portrait painted by Tintoret, who was in no wise afraid of the scoundrel's enmity, although most of the prominent personages of the time quailed before it. Aretino being posed, Tintoret furiously drew a hanger from under his coat. Aretino was terrified lest he should be punished for his malicious tongue, and cried out, "Jacopo, what are you about?" "I am only going to take your measure," said Tintoret complacently; and, measuring him from head to foot, he added, "your height is just two and a half hangers." Aretino's impudence returned. "You're a great madman," he said, "and always up to your pranks." But this grim hint sufficed; the rascal never after dared to slander Tintoret, but, on the contrary, tried to ingratiate himself into his friendship.

In his home Tintoret enjoyed tranquillity. His wife, Faustina de' Vescovi, was thrifty and dignified, and perhaps she was not a little annoyed by the "unpracticalness" of her husband. According to tradition, when he went out she tied up money for him in his handkerchief, and bade him give an exact account of it on his return. Having spent his afternoon and money with congenial spirits at some rendezvous whose name, unlike that of the Mermaid, where Elizabethan wits caroused, has been lost, he playfully assured Madonna Faustina that her allowance had gone to help the poor. She was particular that he should wear the dress of a Venetian citizen; but if he happened to go abroad in rainy weather, she called out to him from an upper window to come back and put on his old clothes. We have glimpses of him passing to and fro in Venice with Marietta, his favorite daughter, a painter of merit, whose early death saddened his later years.[1] Of his other children, two daughters entered a nunnery; a third married Casser, a German; his eldest son, Domenico, adopted his father's profession, and assisted him in his work; another son went to the bad, and was cut off from an inheritance by his father's will. In spite of his habit of giving away pictures, or of charging a small price for them, Tintoret bequeathed a comfortable fortune to his heirs.

A few of his precepts and suggestions concerning art have come down to us through Ridolfi, who had them from Aliense, one of Tintoret's pupils.

"The study of painting is arduous," he used to say; "and to him who advances farthest in it more difficulties appear, the sea grows ever larger."

"Students must never fail to profit by the example of the great masters, Michael Angelo and Titian."

"Nature is always the same; in painting, therefore, muscles must not be varied by caprice."

"In judging a picture, observe if, at the first

---

[1] Marietta was born in 1560, and died in 1590.

examination, the eye is satisfied, and if the author has obeyed the great principles of art; as to the details, each will fall into error. Do not go immediately to look at a new work, but wait till the darts of criticism have all been shot, and men are accustomed to the sight."

Being asked which are the most beautiful colors, he answered, " Black and white; because the former gives force to figures by deepening the shadows, the latter gives the relief."

He insisted that only the experienced artist should draw from living models, which lack, for the most part, grace and symmetrical forms.

" Fine colors," he said, " are sold in the Rialto shops; but design is got from the casket of genius, by hard study and long vigils, and is therefore understood and practiced by but few."

Odoardo Fialeti asked him what to study. " Drawing," replied Tintoret. Somewhat later, Fialeti sought further advice. " Drawing, and again drawing," Tintoret reiterated.

" Art must perfect nature," was his guiding rule; and he instanced that Greek artist who modeled an Aphrodite by selecting the best features of the five most beautiful women he could find.

His studio was in the most retired part of his house. Few were admitted to it, and they had to find their way thither up a dark staircase and along dark passages, by the light of a candle. There he spent most of his time, — a grave man ordinarily, as must ever be the case with genius which ranges the utmost abysses and sublimities; at heart a solitary man, so far as the absence of flesh-and-blood companions constitutes solitude, but forever attended by the great associates of his imagination. Laconic, too, in speech as with his brush; as when, in reply to a long letter from his brother, he wrote simply, "Sir: no." But upon occasion — as that anecdote of Madonna Faustina's allowance shows — he indulged in conviviality; and he had the gift, peculiar to a gentleman, of "being easy with persons of all ranks, and of putting them at ease." "With his friends he preserved great affability. He was copious in fine sayings and witty hits, putting them forth with much grace, but without sign of laughter; and when he deemed it opportune, he knew also how to joke with the great."

Tintoret's genius was only partially acknowledged during his lifetime, and his fame has suffered strange vicissitudes since his death. At times he has been extolled with meaningless extravagance; oftener condemned, after Vasari's lukewarm fashion, or passed over without mention. Not until Mr. Ruskin came and opened the eyes of the world had Tintoret been adequately appreciated for those points of excellence wherein he has neither rival nor second. He has suffered for the same reasons that Shakespeare was long unesteemed in France: his works are bold, very rapid, often unequal, not in the least to be measured by the yardstick of conventionalism; he treats many new subjects, and the old subjects he always treats in new fashion, thereby provoking formalists to accuse him of wilful oddity or caprice; his reputation for swiftness of execution was deemed by many presumptive evidence that he was superficial; above all, his imagination was so rich and so powerful that it required a cognate imagination to follow it.

Moreover, Tintoret was the last master of the great era of Italian painting. After him came schools which did not rely upon originality, but upon the inspiration of former masters. Pictures were but specimens of technique, and the models chosen for imitation were naturally those in which technique could be most easily reduced to rules. The public, as well as the painters themselves, gradually lost the power of valuing art as a *spiritual expression*. Word by word, sentence by sentence, the great language of painting was forgotten, until at last it became as a dead language. It was inevitable that Tintoret's works, which had not always been understood by his contemporaries, should baffle the interpreters of art grammars and the pedagogues of technique.

Again, Tintoret's pigments have suffered more than those of any other master. The darker colors, in many cases, have become almost black; the lighter have faded, and sometimes completely changed.[1] How far this is due to an original defect in the paints, how far to exposure and neglect, I cannot say. It must always be remembered that, as popular canvases have been frequently varnished and restored, many Titians and Raphaels are as fresh to-day as they were when they left the easel. How much remains of the original painting is another question. Directors of galleries aim at pleasing the public, not at respecting the preferences of connoisseurs, and the public craves lively colors. It would feel itself imposed upon if it traveled to Dresden only to find the "Sixtine Madonna" as dark as would probably be the case if the restorer had not interfered. In every gallery you will observe that the crowds flock to the brightest pictures, irrespective of their merits. The fact that they have been kept bright is an advertisement that they are deemed precious; and besides, it requires less time to glance at a clean

---

[1] In some of the paintings at San Giorgio the blues are now milky splotches.

canvas and pass on than to recover, after patient scrutiny and an effort of the imagination, some of the beauty which time and dust conceal. It is significant that the one painting by Tintoret which is most commonly mentioned by all classes of tourists — "St. Mark Freeing a Fugitive Slave" — is precisely that one which the directors of the Venice Academy keep polished as good as new.

I cannot dismiss this subject without alluding to another cause for the slight attention given to Tintoret: his pictures are almost invariably condemned to oblivion by the position in which they have been hung. You must look for them in dark corners near the ceiling, or in cross-lights which render an examination impossible. Of those which still exist in the churches for which they were painted, some have been injured by the drippings from candles; others have been partly hidden by tabernacles, reliquaries, and other objects of church ceremonial. Travelers in Venice a generation ago record that rain leaked through the roof of the School of San Rocco, and soaked some of the canvases; others, hung near windows, have had to suffer from the strong sunlight for centuries. In the Ducal Palace, one series of ceiling paintings have succumbed to the daubing of restorers, and are now hardly recognizable as being Tintoret's; while the matchless "Paradise," when I recently saw it,[1] was falling rapidly to decay. The seams where the vast canvas was originally joined had rotted in many places; the canvas itself was warped and rumpled, forming little shelves and unevennesses on which the dust had collected so as to hide the colors; and from the ceiling dangled a ragged fringe of cobwebs, in some places two or three feet long.

A few generations hence, when these incomparable works have been irretrievably damaged, posterity will wonder — with a wonder intensified by indignation — that we allowed them to perish. Early Christians, who mutilated pagan works of art because they believed them to be pernicious, may be excused; but what excuse has our age to offer? We pretend to cherish all manifestations of culture, and we have ample means to preserve them; yet whilst our museums are daily adding to their collections of half-barbarous antiquities, dug up in Arizona, in Mexico, in Yucatan, in Peru, in Asia Minor, in Mesopotamia, there are surely hastening to destruction scores of the works of the mightiest genius who ever honored painting. During the past twenty years, New York millionaires have paid more for the immoralities and inanities of modern French painters than would be necessary to erect a separate gallery in Venice for

[1] In August, 1889.

the proper preservation of Tintoret's masterpieces. If there were but a single manuscript of *Hamlet* in the world, and no printing-presses, what should we say to those who allowed it to perish through neglect? Yet there are many of Tintoret's pictures, each of them as precious in its way as a page of *Hamlet*, which we raise no voice to save. In our selfishness, we forget that the treasures which we have inherited from the past are not ours to dissipate and destroy; we hold them in trust for the future, and woe unto us if, unmindful of our responsibility, we prove careless stewards.[1]

II. HIS WORKS.

What, then, are some of the qualities of Tintoret's genius? First of all, he had vast scope: Christian and classic lore, the legend and story of Venice, contemporary scenes, and portraiture, —

[1] So long as the originals exist, copies of great paintings are as unsatisfactory as a Beethoven symphony or a Wagner opera on the piano; but when the originals have perished, copies may serve a worthy purpose in perpetuating at least the concept and general treatment of the painter. It is greatly to be desired that some capable student should do for Tintoret what Toschi has done for Correggio at Parma. A series of faithfully executed sketches would enable posterity to judge of Tintoret's range of imagination and inexhaustible powers of treatment, although his coloring and drawing could not be reproduced. Many of his paintings have never been engraved, and not one has been well engraved.

all these lay within his province. But scope alone, unguided by rarer powers, does not suffice for the equipment of the supreme master. Rubens had scope, even Doré had it, and neither ranks among the foremost. In Tintoret it was accompanied by a most intense imagination, which penetrated to the elemental reality and understood the intertangled relations of life. Imagination operated through him with a vigor more like Nature's own than that of any other man except Shakespeare; a vigor which seems at once inexhaustible and effortless, which never wastes and never scants. In creating a beggar or a seraph he expended just as much energy as was necessary for each; you do not feel that one was harder for him than the other. Tintoret's creations have this further resemblance to Shakespeare's: *they live!* You do not exclaim, "This is a great picture!" but, "This is a great scene!" He is like a traveler who brings back views from a strange country: albeit you have never been there, yet the views are so real, the figures are painted so freely and lifelike, and not in conscious or conventional attitudes, that you cannot doubt their faithfulness, and are absorbed by the wonders and beauties they present.

Tintoret never conspires to startle you by sensational devices. Even in those works where he is most daring he is really painting what his imagination saw naturally, and is no more bent on inventing oddities and marvels than was John in the *Apocalypse*. Before beginning a Biblical or an historical subject, he seems to have asked himself, " How did this look to a bystander ? " and he relies upon the actuality of the scene to produce the desired impression. He has been charged, sometimes, with making Christ and his disciples too vulgar. Other painters have so accustomed you to look for a kingly personage in Christ, and for princely garments on his followers, that when you first see a " Last Supper " by Tintoret you miss the habitual elegance; for he shows you simple and earnest but not ignoble fishermen and artisans of Judea. If you contemplate them wisely, your astonishment will deepen as you reflect that it was through and by such lowly and zealous men as these, and not by philosophers and prelates and princes, that the gospel of brotherly love was disseminated among mankind. It is legitimate for an artist to invest an historic character with emblems which bespeak the significance posterity has attached to him; but it is wholesome to see him as he probably appeared to his contemporaries, before subsequent generations have discovered a retroactive importance in his career. Tintoret employed now one method and now the other, and whosoever has been moved by the "Christ before Pilate" and "The Crucifixion" of the School of San Rocco needs not to be told that pathos and sublimity belong only to the former method.

Tintoret's versatility would have made a lesser man renowned. He counted it but an amusement, when the learned critics chided him for not obeying academic rules, to imitate the style of Titian, or Paul Veronese, or Schiavone, so that the critics themselves were deceived and confounded. He invariably adapted his treatment to the requirements of each work: if it was to be viewed from a considerable distance, he painted broadly; if it was to be seen near, no one surpassed him in the delicacy and carefulness of his finish. This sense of fitness governed his composition as well as his drawing. In a picture intended for a refectory, for instance, he introduced proportions in harmony with the dimensions of that refectory, causing it to appear more spacious and imposing. Where Tintoret's figures are not correctly drawn, the apparent fault was often intentional: restore the picture to the position for which he designed it, and the drawing will no longer offend; for he always took into account the distance and angle from which the spectator would look, and he is not responsible for the changes in location. In studying any picture, remember that there is one, and only one, point of view where it can be seen as the artist wished it to be seen. If you stand too far or too near, you will miss his purpose. In a portrait by Titian or Tintoret, no line, no dot of color is superfluous: you must adjust your vision until the tiniest flake of white on the tip of the chin or on the pupils of the eyes shows you its reason for being there. Try to imagine that last perfecting touch away, and you will learn its value. For these men did nothing haphazard: they would as soon have wasted diamonds and rubies as their precious colors; every hair of their pencil was a nerve through which their imagination transmitted itself to the canvas.

Although it be well-nigh impossible to describe a painting so that one who has not seen it can derive profit from the description, I shall attempt to point out a few of the characteristics of some of Tintoret's other works, in the hope of refreshing the memory of readers who are already familiar with them, and of stimulating the interest of those who may see them hereafter. It is the *thought* Tintoret has expressed, and not the technique of his manner, to which I would call attention, believing that this can be in some measure made real even to those who cannot refer to the paintings themselves.

One fact impresses us immediately, — Tintoret's

originality. Previous painters had used all the familiar Christian themes so often, that there had grown up a conventional form of representing each; but, although Tintoret used these themes, his treatment of them rarely recalls that of any other painters, and always demands fresh study. Giotto may be said to have fixed the norm which his successors generally followed, diverging from it only in details. Tintoret established a new norm. Moreover, he never copied himself; his inexhaustible imagination refused to repeat. It represented the same subject under different aspects, never twice alike. We have many replicas of Raphael's and Titian's works, but none, so far as I know, of Tintoret's. In rare cases where two copies of a painting by him exist, one is the sketch.

In one famous instance he is brought into direct comparison with his rival, Titian. They both painted "The Presentation of the Virgin," in somewhat similar manner. Titian conceives the scene as follows: In front of a stately pile of buildings, two flights of steps lead up to the threshold of the Temple, where stands a venerable high priest; near him are two other ecclesiastics and a youth. Spectators look out from the windows and balconies of the adjoining edifice upon Mary, a pretty little maiden, who has reached the first step of the second staircase, and, looking up at the high priest, prepares to finish the ascent. Immediately back of her figure is an ornate Corinthian column. Her mother and a friend wait at the foot of the staircase, and a goodly company of Venetian nobles is gathered near them, — like pleasure-seekers taking a stroll, who stop for a moment to witness a chance episode. An old woman with a basket of eggs sits in the foreground. A colonnade and pyramid close in the picture on the left,[1] and a pleasing view of mountains stretches out behind.

This is Tintoret's conception: A high priest, patriarchal in dignity, stands at the top of a flight of steps leading to the door of the Temple. Just below him Mary is mounting, her slight form and dress being beautifully contrasted with the sky beyond. Behind her is a young woman (probably her mother, Anne) carrying a young child. At the foot of the steps, in the centre of the painting, another mother (one of Tintoret's matchless creations) is pointing toward Mary, and telling her little daughter that she, too, will erelong be presented at the Temple. Two girls recline on the steps near by. On the left, seven or eight old men and idlers (such as one still sees at the approach to churches in Italy, and to mosques and synagogues in the Orient) are ranged along the stairs, indolently

---

[1] I use *left* and *right* to denote the positions as the spectator faces the picture.

watching the scene. The shadow of the building falls upon them, and prevents their figures from being too prominent. There is no suggestion of Venice or Venetian nobles. The attention is not distracted by costly apparel or imposing architecture, but is fixed upon the chief actors,— upon the venerableness of the high priest, the simplicity and confidingness of the little maiden, and the magnificent forms and naturalness of the women.

Critics have disputed whether Titian's picture or Tintoret's be the earlier. The presumption is in favor of the former,[1] but there is no reason to cry plagiarism against either, because each master has worked out a similar conception with characteristic independence. The central idea — the youthful Virgin ascending the steps of the Temple to be received by the high priest — may be seen in one of Giotto's frescoes.[2] What we admire is the originality of treatment in both pictures. To me, Tintoret's conception seems the more nobly appropriate; and I know not in which of Titian's works to look for a counterpart of that woman in Tintoret's foreground, so easy, so living, so superb.

---

[1] Crowe and Cavalcaselle give 1539 as the date of Titian's "Presentation;" 1545-46 is usually assigned as the date of Tintoret's.

[2] At the Arena, Padua.

As an example of Tintoret's insight into the spiritual world, turn to his picture of Lucifer.[1] From early Christian times, the Evil One has been represented by very crude and vulgar symbols. A hideous face, horns, a tail, and cloven hoofs have come to be his accepted signs. Such a monster could never tempt even the frailest striver after righteousness; for this conception illustrates the loathsomeness of the *results* of sin, and not the allurements by which sin entraps us. It would be equally appropriate to show to a lover a crumbling skeleton as the effigy of the woman whom he loves. The Devil would make no converts if he announced himself to be the Devil, and dangled before men's eyes the despair, the degradation, the infinite remorse, which are his actual merchandise, instead of the fleeting pleasures and deceitful promises under which he masks them. He is no bungler or fool, but supremely skilful in proportioning his enticement to the strength of his victim, and very alert in choosing the moment most favorable for attack. Goethe, in his Mephistopheles, has portrayed the enemy of good under one of his aspects, emphasizing the cynical and wicked rather than the seductive and plausible qualities. Tintoret has depicted the latter. His Lucifer is still an angel, though fallen. He has a commanding and

[1] At the School of San Rocco, Venice.

beautiful form, and a countenance which at first fascinates, until, on searching it more deeply, you fancy you discern a suggestion of duplicity, a hint of sensuality, in it. Bright-hued and strong are the plumes of his wings, and a circlet of jewels sparkles on his left arm, the sole emblem of the wearer's wealth. Here is indeed a being whose beauty might seduce, whose guile might deceive, — one whose presence dazzles and attracts, for it has majesty and grace and charm. Here is a fit embodiment of that ambition which shrinks not from crime in order to possess power; or of that false pleasure which decoys men from duty, and, still flying beyond reach, leads its prisoner deeper and deeper into the abominations of the abyss.

With equal originality and truth, Tintoret has illustrated the allegory of the temptation of St. Anthony.[1] This subject is usually treated either absurdly or grotesquely; as when the saint is discovered in a grotto through which bats, mice, witches, and imps flit and gambol. Not one of these ridiculous creatures, we may safely say, would frighten or tempt anybody. But who are the enemies that a man whose life is dedicated to holiness, and who has taken the three vows of poverty, chastity, and obedience, must resist?

[1] In the church of San Trovaso, Venice.

Tintoret's picture gives the answer. In it one of the figures, typifying Riches, offers gold and precious gems. "Why live a beggar?" she pleads softly; "take these and have power." A second figure, Voluptuousness, is that of a woman fair in body. "Come with me," she urges; "let us taste of joy together while there is still time." A third, who (I think) represents Unbelief or Heresy, has already dashed the saint's missal and rosary to the ground, has snatched up his scourge, and, endeavoring to drag him away, has plucked off his mantle. "Come with me," this tempter seems to say; "there will be no more scourging, and fasting, and mortification; with me your life shall be without care and unrestrained." Nevertheless, Anthony, thus hard beset, looks heavenward, uttering a prayer for succor. Are not these apt personifications of those lower impulses to which even men of high resolve have succumbed? All the witches of the Brocken and all the bats in a Pharaoh's tomb have nothing alluring about them.

There are few of Tintoret's paintings which will not make similar revelations, if you look attentively. Often what appears to be only a casual accessory is the key to the whole composition. Let me cite two instances of his imaginative use of color. The first occurs in "The Martyrdom of St. Stephen."[1] The saint has fallen on his knees beneath the stoning of his persecutors, but there is no melodramatic spurting of blood or sign of physical pain. His face betokens fortitude, resignation, and forgiveness of his tormentors. He gazes up steadfastly into heaven, and sees the glory of God, and Jesus standing on the right hand of God. The Almighty is clothed in a robe of red and a black mantle. In the background on earth, behind the martyr, a crowd watch the persecution; they are too far away for us to distinguish faces, but one of them, who is seated, is clothed in black and red. It is Paul, soon to acknowledge Christ and put on the livery of God. Again, in the "Paradise," Tintoret gives profound significance to color as a symbol: Moses, the witness to the Old Covenant, and Christ, the witness to the New Covenant, have robes of similar colors.

The Doges' Palace contains a score of Tintoret's imaginative paintings and many of his portraits, and there are few churches in Venice which have not at least one altar-piece by him. His best portraits, as I think, outrank even Titian's best: they have a vital quality, an *inevitableness*, which can be felt, but not described. What a concourse of doges, senators, procurators, nobles, and

---

[1] In the church of San Giorgio Maggiore, Venice. Mr. Ruskin was the first to point out this stroke of genius.

soldiers he has portrayed! Their grave, refined faces, their stately carriage, the sobriety as often as the sumptuousness of their dress, bear witness to the glory and power of Venice; that glory and power which had begun to decline in the sixteenth century, though the Venetians perceived it not. They misread the signs. They could not believe that Venice, which had continually grown in wealth during ten centuries, could decline or perish. *Esto perpetua!* — may she live forever! — was the last prayer of her historian, Sarpi, the abiding dream of all her citizens.

It was Tintoret's pride to immortalize on canvas her legends and her history, and to illustrate her grandeur by means of allegory. He painted the popular stories of the recovery of St. Mark's body from Alexandria, and of the miracles performed by that holy patron. He painted the siege of Zara, the battle of Lepanto, and the ambassadors of Venice holding head before the haughtiness of Frederick Babarossa. He painted Venice enthroned among the gods, and Venice as mistress of the sea.

But his genius was not confined to the expression of pomp and patriotism. It delighted not only in majestic flights of imagination, but also in contemplating and in setting forth pure beauty. In one of the smaller rooms of the Ducal Palace are two classic subjects by him, — "Mercury and the Graces," "Ariadne and Bacchus," — which, whether we regard their perfect symmetry, or the grace of their forms, or the delicious poetic spirit that emanates from them like fragrance from a bed of lilies, have few rivals in loveliness. They arouse in some beholders a mood akin to that which a joyous theme in one of Beethoven's symphonies can arouse, — a mood sweeter than hope itself, or the brightest afterglow of memory; for, while it lasts, the present, flooded with peace and beauty and a nameless ecstacy, satisfies the soul.

The School of San Rocco possesses sixty-four pictures by Tintoret. This series, illustrating the principal events in the Old and New Testaments, is quite without parallel, not only in extent, but in the excellence of a large number of the separate paintings. You pass from one to another as from scene to scene in Shakespeare; and it is only when you return to the works of lesser men that you realize the richness and strength of the master, who has lifted you to his level so easily that you were conscious of no effort. The halls in which these paintings are kept are utterly inadequate for their proper examination: not one can be seen in a favorable light; many are almost buried in gloom, or hidden in the equally impenetrable glare that falls on their surface from the cross-lights of conflicting windows. Some of the canvases have been injured by water; the colors have grown dim or dingy with age; and in some cases "restorers"[1] have blurred the outlines and brought discord among the tones. Nevertheless, who that has once seen can ever forget many of those paintings? The original conception looms up beautiful and grand from amid the wreck of time and neglect, like a mutilated, earth-stained Greek statue, and your imagination exerts itself to see the work as it must have appeared when the colors were fresh. Who can forget that flock of angels in "The Annunciation;" or "The Visit of the Magi;" or "The Flight into Egypt;" or the terrible "Slaughter of the Innocents," which seems to have been painted in blood, though there is hardly any blood to be seen; or "The Adoration of the Shepherds;" or "Christ's Agony in Gethsemane;" or "Christ before Pilate;" or "Christ being led to Calvary"?

The series concludes with "The Crucifixion," a masterpiece before which artists and amateurs, and even academic critics, have stood in mute wonder. It is a panoramic summary of the last acts in the

---

[1] One painting bears the inscription, REST. ANTONIVS FLORIAN, 1834. "Exactly in proportion to a man's idiocy," Mr. Ruskin remarks, "is always the size of the letters in which he writes his name on the picture that he spoils."

persecution of Christ. No detail which the Evangelists furnish has been omitted, but all details have been subordinated to a unity so vast and impressive that it eludes analysis. Primarily, this is a pictorial representation of an historical event; but for the Christian believer it is an image of the profoundest religious meaning. There are many groups, but if you study each group you will discover that without it something would have been wanting to the whole. Here are Romans, to whom the spectacle has no moral interest; they are soldiers and judges, executing unperturbed the Roman law upon the person of a Jew who has stirred up the wrath of his fellows and caused a popular tumult. Here are Jews, mocking and full of hate. Here, too, is the little remnant of Jews who, believing in the victim as their master, are faithful to him unto death. Is not the indifference or the idle curiosity of some of the spectators as significant as the cruelty of his enemies and the devotion and anguish of his friends? For consider well what it implies that any human being should gaze unmoved, or moved only as by an every-day occurrence, at a fellow creature suffering the penalty of death. Is life, then, so cheap? Is a human soul of so slight account that men can cast lots or jeer while it passes in agony from earth forever? Who can estimate the cruelty which delights in the torments of that struggle? And if this sacrifice be viewed with the eyes of a Christian, and not of an impassive observer, if the victim be esteemed not merely a man, but the Son of God, what words shall describe its solemnity?

Tintoret has painted all this into his picture, in which the central object is the cross with Christ upon it. His head has sunk upon his bosom, and we imagine that with his downcast eyes he beholds the group of holy women at the foot of the cross, and says to Mary, " Woman, behold thy son." That group is the most pathetic that painter ever drew. Some of the women, overwhelmed by grief, have fainted. Not by their faces, but by their drooping, motionless bodies, can you infer the unspeakable burden which is crushing them. One kneels; another — Magdalen, it may be — has risen, and looks up at the expiring Saviour. A venerable disciple gazes tenderly at the face of the Virgin, who has swooned. A younger disciple lifts his eyes toward Christ. They cannot help; they cannot speak; they can only wait and sorrow. Who shall utter the agony that love feels when it is powerless to relieve the suffering of its beloved!

Behind this group stands a man holding a bowl, into which another man, who has climbed a ladder resting against the back of the cross, dips a sponge stuck on a spear. At the left, other executioners are raising the cross on which one of the malefactors has been bound. Some men in front are tugging at ropes; others behind are pushing or steadying it. Hammers, adzes, a saw, and other tools bestrew the ground. Farther on are many spectators, — a Roman officer in armor, elders, dignitaries, and a soldier bearing the Roman standard. Some point toward Christ, and evidently say to one another: "That is the impostor who calls himself the Son of God and the King of the Jews. Where is his pretended might?" A little in the background, a mounted spearman has thrown the reins on the neck of his ass, which complacently feeds on withered palm leaves, — an imaginative touch characteristic of Tintoret, which will not be lost on those who recall Christ's entry into Jerusalem a few days before.

In the foreground, to the right, a man is digging a hole for the cross of the second malefactor, while soldiers are drawing lots for Christ's garments, and other mounted soldiers are watching the proceedings near by. A little beyond, another group is busy attaching that malefactor to his cross; one boring a hole for the spike to pierce his hand, another holding down his legs so that they can be bound, while a third has a rope. In the distance, men hurry toward the scene, lest they be too late to enjoy it; and the foremost camels of a caravan on its way into the city appear just at a turn in the road. For traffic and the daily toil of men are not interrupted by the crucifixion of Christ, though soldiers and idlers have come out to witness it. On the left there is a palace, and then hills succeeded by craggy mountains. The clouds have deepened almost into darkness along the horizon. The sun, as it sinks into this gloom, appears as a huge disk of ghastly light, and this disk forms a dim halo behind Christ's head. Yet a little while and the earth shall be wholly darkened, and these curious, careless spectators shall flee away in terror.[1]

Such, told briefly and inadequately, — for language can only hint at the effects of painting, — is this solemn event as conceived by Tintoret's imagination.[2]

We have no evidence that Tintoret visited Rome, nor any record of his journeys, except that to Mantua, yet we may be sure that he was familiar with the scenery of the mainland. The woods and foliage, the streams, valleys, and meadows,

---

[1] In a great picture, now ruined, at the abandoned Bavarian palace of Schleissheim, near Munich, Tintoret has represented the Crucifixion in its later aspect.

[2] This is one of the four or five paintings which Tintoret signed. It was finished in 1565. His receipt for its payment still exists. It is dated March 9, 1566. The sum received was two hundred and fifty ducats.

the little hills and picturesque mountains, which abound in his paintings, he did not see at Venice. Our lack of information leaves us in doubt, therefore, whether he studied Michael Angelo's " Last Judgment" in the Sixtine Chapel. If he never went to Rome, he probably was acquainted, from engravings or copies, with the composition of that extraordinary work; yet his own painting of that subject bears so little resemblance to Michael Angelo's that it seems to have been produced independently.

The masterpiece of the Sixtine Chapel is so complicated that it bewilders the student, until he observes that the principal groups are roughly arranged in an immense irregular horseshoe, the points of which are near the bottom of the lower wall, while Christ, the chief figure, is inclosed in the upper oval. Four fifths of the action takes place in the air, the lower portion alone of the fresco being occupied by the river Styx and its adjacent bank. In its present nearly ruined condition we cannot guess the original effect of this work; but I doubt whether it could ever have satisfied the beholder's instinctive demand for harmony. The groups, even the individuals, seem isolated, not only in space but in spirit. There is not, nor could there be, a single prevailing passion. The only characteristic which applies to the whole work is tremendous energy. Whatever of agony, of fury, of stubbornness, of determination, can be expressed by the human body, is expressed here. There is no muscle or tendon which is not exhibited in various positions; no posture of limbs or trunk which is not represented. The resurrection of the *body* is illustrated in a hundred ways, and the expression of the faces is of secondary importance. Here, patriarchs have the vigor of Titans; saints are as robust as athletes; Christ himself might be a majestically stern Apollo. Not without reason may we call these effigies of restless, writhing human beings wonderful diagrams of anatomy and concrete illustrations of dynamics. Even the saved, who occupy the higher regions, are not tranquil. In striving to comprehend these whirlwinds of action, the mind is wearied and baffled. Unit by unit you examine this multitude, and you are amazed in turn by sublimity, or horror, or power.

The space [1] to which Tintoret had to adapt his picture of "The Last Judgment" is oblong, about fifty feet high and twenty feet broad. In the upper part of the heavens Christ is represented, not in the character of the inexorable Judge, but in that of the Shepherd who welcomes his faithful flock to Paradise; for the resurrection and judg-

[1] In the church of Santa Maria dell' Orto, Venice.

ment are coincident. On one side, near Christ, John the Baptist is kneeling, and Mary and the repentant sinner, who bears a cross, are near; on the other side are personifications of the cardinal virtues. Extremely lovely is Charity, carrying in her arms two young children to present to the Saviour. Zones of fleecy clouds separate the upper part of the painting into sections, in which the saints are ranked; but the distribution seems natural, not arbitrary, and serves to prevent confusion among so many figures. Midway in the scene, angels plunge earthward to rouse the dead. Michael, with his terrible sword unsheathed, pursues the wicked toward a mighty river, which sweeps irresistibly into the abyss. In the distance, on a low shelf of sand amid the waters, is huddled a crowd of sinners, too indolent or too terrified to struggle against the flood which must soon engulf them. Crouching, they await their doom. In them Tintoret has perhaps typified those miserable creatures whom Dante describes as "*a Dio spiacenti ed a' nemici sui*," — hateful to God and to his enemies. Demons convoy a bark-load of the damned through the hellish torrent. And on the shore what a spectacle! Bodies starting from their graves, some not yet clothed with flesh, some with leafy branches growing from their arms, some striving to free themselves from the earth into which corruption resolved them; everywhere signs of the suddenness and awfulness of that supreme moment when the dead shall rise again in the forms they bore when alive, and go to the eternal abode, of bliss or punishment, for which each has fitted himself by his career on earth.

A parallel has frequently been drawn between the genius of Michael Angelo and that of Dante, and many have deplored the loss of that portfolio in which Michael Angelo is known to have made a series of illustrations to *The Divine Comedy*. The resemblance between the supreme Tuscan poet and the supreme Tuscan artist seems to me, however, to hold only when we limit our view to Dante as the author of the *Inferno*. In energy, in intense perception of evil, in unswerving condemnation of sin, in austerity, in appreciation of the terror of life, the poet and the painter were indeed akin. These are the characteristics which most readers associate with Dante's genius, for the reason that most readers go no farther than the *Inferno*, or are unable to comprehend the more spiritual sublimity of the *Purgatorio* and the *Paradiso*. The *Inferno* describes torments which the most sluggish person can understand, and the contrasts of lurid flames and impenetrable gloom by which the scenes in hell are diversified are so vivid as to require no commentary. We marvel at the imagination that could traverse unparalyzed these horrors and dare to report them. But Dante's genius stopped not here: it passed in review all human nature, from its lowest sinful condition to that highest excellence when it merges with God. Though Evil be a terrible reality, Dante saw that Love is even more real, the source and the goal of all things, and he proved his universality by his power to describe it. And they whose imagination is strong enough to follow him through the regions of the blessed incline to rank the third canticle of his "sacred poem" even higher than the first.

Among painters, Tintoret only has, like Dante, swept through the full circuit of human experience and aspiration. He has shown us the anguish of the damned in his "Last Judgment," and the peace and bliss of the blessed in his "Paradise." That "The Last Judgment" should be Michael Angelo's masterpiece, and that he should have painted it on the altar wall of the Pope's favorite chapel, are fatally appropriate. In that terrific scene, the judge is not Christ, but Michael Angelo himself; a righteous man, who looked out upon the iniquities of his time and dared to condemn them; a religious man, who, coming to Rome, the religious centre of Christendom, discovered there a second Sodom, in which pope, cardinals, and bishops were the most shameless offenders; a patriotic man, who had fought for the liberty of his beloved Florence, and had beheld her, through the treachery of some and the apathy of others, become the slave of a corrupt master. No wonder that the terror and anguish, the depravity and hopelessness, of life should have eaten into Michael Angelo's soul. As he worked solitarily in the Sixtine Chapel, no wonder that a vision of the retribution which shall overtake the wicked should have possessed his imagination, and transformed the artist into the judge. Day by day, a spirit mightier than theirs painted the protest which Savonarola, Zwingli, Luther, and Calvin had preached, — the spirit of a Job united to that of an Isaiah.

Not less appropriate was it that the genius of Tintoret and of Venetian art should culminate in the representation of Paradise. Of all commonwealths, Venice had enjoyed the longest prosperity; of all peoples, hers had been the most sensitive to the joy of life. Even at the end of the sixteenth century, when her power abroad had been curtailed, and when luxury at home was slowly enervating the integrity of her citizens, she was still outwardly imposing, magnificent. No pope had ever succeeded, either by guile or by force, in ravishing her independence. Her immemorial glory blazed across the past and irradiated the present, as the setting sun spreads an avenue of splendor upon the ocean and fills the heavens with golden and purple light. Venice was indeed the abode of Joy; and Tintoret, at the close of a long career, in which he had witnessed all the aspects and pondered all the possibilities of human life, was filled, like Dante, with hope, and felt Joy and Love to be the supreme realities, the everlasting fulfilments, of mankind's desires.

If the Last Judgment is an "unimaginable" theme, as Mr. Ruskin remarks, how much more so is Paradise! Men have always found it easier to represent grief than happiness, villainy than virtue, shadows than sunshine; for the former are by their nature limited, and draw their own outlines, while the latter have a quality of boundlessness which to define abridges it. Moreover, pleasure is oftenest unconscious, and always individual; pain, on the contrary, is too conscious of self, and is manifest in attributes common to many. Nevertheless, Tintoret has achieved the seeming impossibility of representing, so far as painting may, the happiness, unmixed and eternal, of the celestial host.

His painting is known to most visitors at Venice as being the largest in the world. The ordinary traveler, after reading the dimensions in his guidebook, looks up at the canvas, and sees crowds of figures and colors grown dark; wonders what it all means, and why the janitor does not sweep down the dust and cobwebs; and then turns away to devote equal attention to the black panel where Marino Faliero's portrait would be had he not died a traitor's death. In like manner, I have seen intelligent strangers exhaust the treasures of the Acropolis of Athens in a quarter of an hour, and return to their hotel to read the last English newspaper. But let him who would commune with one of the few supreme masterpieces of art sit down patiently and reverently before Tintoret's "Paradise," and he will be rewarded by revelations proportioned to his study. As soon as his eyes grow used to the dimness of the hall, the tones of the canvas begin to be intelligible to him: it is as if he heard a symphony played in a lower key than the composer intended; many of the original effects are lost, but harmony interpenetrates and unifies all the parts. When he has adjusted his eyes to this pitch, he can examine the figures separately; until, little by little, in what seemed a vast confused multitude, he will be aware of the presence of an all-controlling order; and he will gaze at last understandingly, as in a vision, upon the congregations of heaven as they are unfolded in Tintoret's design.

Christ is seated in the central upper part of the painting: his left hand rests on a crystal globe; innumerable rays of light illumine his head and dart in all directions. Opposite to him is the Madonna, above whom sparkles a circlet of stars. At Christ's left soars the archangel Michael bearing the heavenly scales; at Mary's right is Gabriel with a spray of lilies. A cloud of countless cherubs hovers at the feet of the Divine Personage; while on each side of the archangels, curving toward the upper extremities of the canvas, sweep companies of seraphim and cherubim, and the thrones, principalities, and powers, and angels with swords, sceptres, and globes. These form the first circle of the angelic host, who from eternity have held their station nearest to their Lord. Below them is a larger circle, composed of those spirits who, by prophecy or preaching, established and extended the kingdom of God on earth. On the left we see the forerunners of Christ, — David playing the cithern, Moses holding up the tables of the law, Noah with his ark, Solomon, Abraham, and the other patriarchs; and near these we distinguish John the Baptist, who displays a scroll on which is written *Ecce Agnus*. Midway in this circle are the Evangelists, the four corners of the Christian temple, and the intermediaries between the old and new dispensations. Here is Mark accompanied by his lion, Luke and his ox, Matthew with pen in hand, and John with his book resting on an eagle. As the line sweeps on, we see the early fathers, doctors, and great popes, — Peter and Gregory; Paul, the apostle militant, recognizable by his sword; Jerome, Ambrose, and Augustine. In the centre, between Luke and Matthew, is the third archangel, Raphael, whose clasped hands and upturned face betoken a soul rapt in adoration. The third and lowest circle is made up of many groups of martyrs and holy men and women, the great body of the Church of Christ. Among the throng on the left are Barbara; Catherine with her wheel; Francis of Assisi and Dominick, the founders of the great religious orders; Giustina bearing a palm branch; St. George (with banner), Lawrence, Sebastian, Agnes, and Stephen, each recognizable by a familiar emblem. In the centre, along the bottom of the painting, hover clusters of worshiping angels; beyond them, more saints, Monica, and Magdalen; then Rachel and a troop of lovely children, and Christopher, who carried the boy Christ on his shoulder here below, and now carries a globe. At last, on the extreme right, we reach the assembly of prelates and theologians.

With this key to the general distribution, the student who has Tintoret's " Paradise " before him can recognize scores of other figures. He will compare Tintoret's portrayal of each saint, or prophet, or martyr with conceptions other painters have drawn; and if he reflect that any one of these groups, and many of these figures singly, would have sufficed to establish the renown of an artist less masterly than Tintoret, his astonishment will swell into admiration, and this into awe, when he surveys the work as a whole. Who can describe the effect of the innumerable multitude? Cast your eyes almost anywhere upon the canvas, and lo! out of the deeper, distant spaces angelic countenances loom up. Forms, though distinctly outlined, by some magic seem diaphanous; and the farther your gaze penetrates, the brighter is the light which radiates throughout heaven from the throne of Christ. Still more marvelous is the sense of infinite tranquillity, even in those figures which are moving. These are veritable spirits, though they have human bodies, and they move or rest with equal ease. In this heavenly ether there is no effort. Even those rushing seraphim, whose majestic pinions seem to beat melody from air in their rhythmic flight, suggest a certain grand repose begotten of motion itself, — a repose akin to that produced by the sight of the sea, whose myriad little waves dance and glisten, or of Niagara, whose falling flood seems stationary. The spectator who has risen to this conception will not fail to note the light of a joy, not vehement but profound, which bathes every face; and how the action of every individual and of every group is in some manner addressed to Christ, and would be incomplete but for that divine centre. Christ and the Madonna, and the dove of the Holy Spirit floating between them, he will look at first and turn from last, — the noblest personification of ideal manhood and ideal womanhood that ever painter expressed. The embodiment and essence of *Love*, which is the author of all good, they are enthroned amid the serenity of the highest heaven. Round them wheels the inner circle of the archangels and the angels, the symbols of divine *Power*. Then, in ever-widening circles, the saints and apostles and prophets, and the elect of every clime and condition, all children of *Faith* and exemplars of *Charity*, float and revolve in bliss forevermore. And it needs no strain of the imagination to hear the hosannas which the morning stars sing together, and all the sons of God shout for joy.[1]

The dark chapel of the Rucellai, in the church of Santa Maria Novella in Florence, has a dingy

---

[1] In the execution of the "Paradise" he was assisted by his son Domenico. If Tintoret was born in 1512, most of the work was done after his eightieth year, an indication of physical vigor

altar-piece representing the Virgin and the infant Christ. Cimabue painted it; and when it was finished the Florentines made a holiday, and bore the picture through the streets, amid great rejoicing, to the chapel where it now hangs. That stiff and awkward Madonna, that doll-like Child, were hailed by them as the highest achievement of painting. For us Cimabue's masterpiece has only an historic interest, — we find no charm in its Byzantine rigidness. Yet that crude work was the seed of Italian painting, and if we follow its growth during three centuries we shall be led to the "Paradise" of Tintoret, in which are embodied all the excellences and advances of the painter's art. Between that humble beginning and that glorious culmination an army of artists and myriads of paintings intervene. If we look deep enough, we shall be conscious that they were all agents whereby a mighty spirit was seeking to express itself to man, — a spirit which first appealed to human piety through the symbols of religion, and which, as its agents acquired skill and reach, bodied itself forth in higher images and in conscious forms. The name of that spirit is Beauty, never to be found perfect in the outer world, but

almost unparalleled. A rapid study for another "Paradise," in which the groups are arranged on a different plan, reminding one of Dante's description of the Celestial Rose, is now in the Louvre.

known as it communicates through the senses portents of itself which the soul sublimes into that ideal unity by which the laws of nature and the destiny of man are beheld in their highest aspect. True Worship, as in the sweet piety of Fra Angelico, led to Beauty; to Beauty also, along an inevitable path, led the pursuit of Truth by the sixteenth century masters, latest among whom was Tintoret: for Beauty is the final seal and test of both Holiness and Truth.

GIORDANO BRUNO: HIS TRIAL, OPINIONS, AND DEATH[1]

I

On Saturday, the 23d of May, 1592, Giovanni Mocenigo, son of the late excellent Marcantonio Mocenigo, addressed to the Father Inquisitor of Venice a letter containing charges of heresy against Giordano Bruno, the Nolan. Among other things, he alleged that Bruno had said "that it is a great blasphemy to say, as Catholics do, that bread is changed to flesh; that he is hostile to the mass; that no religion satisfies him; that Christ was a good-for-nothing, and did wretched tricks to seduce the people, and ought to have been hanged; that there is no separating God into persons; that the world is eternal; that worlds are infinite, and God makes an infinite number of them continually; that Christ wrought apparent miracles and was a magician, and so were the Apostles; that Christ showed that he died unwillingly, and evaded death as long as he could; that there is no punishment of sins; and that souls

[1] First printed in *The Atlantic Monthly*, March, 1890.

created by the agency of nature pass from one animal into another; and that as the brutes are begotten of corruption, so also are men. Further, he has denied that the Virgin could have borne a child; he asserted that our Catholic faith is full of blasphemies against the majesty of God; that he wished to give himself to the diviner's art, and draw the whole world after him; that St. Thomas and all the doctors were blockheads compared with himself. Therefore, urged by my conscience and by command of my confessor, I have denounced this Bruno to the Holy Office. Suspecting that he might depart, I have locked him up in one of my rooms, at your requisition; and because I believe him possessed of a demon, I pray you to take speedy resolution concerning him."

Two days later, this Mocenigo, of whom we know no more than that he belonged to one of the illustrious families of Venice, and was thirty-four years of age, added to his accusations: "On that day when I had Giordano Bruno locked up, on my asking him if he would teach me what he had promised, in view of the many courtesies and gifts he had had from me, so that I might not accuse him of the many wicked words which he had said to me, both against our Lord and against the Holy Catholic Church, he replied that he was not afraid of the Inquisition, because he offended nobody in living as he chose ; and then that he did not remember to have said anything bad to me, and that even if he had said it he had said it to me alone, and that he did not fear that I could harm him in this way, and that, even should he come under the hand of the Inquisition, it could at the most force him to wear his friar's gown again."

On May 29, Mocenigo, who had in the mean time, at the suggestion of the Inquisition, dredged in the slimy depths of his memory for other charges, informed the Father Inquisitor that he had heard Bruno say " that the forms which the Church now uses are not those which the Apostles used, because the Apostles, by preaching and by example of a good life, converted the people, but that now he who will not be a Catholic must suffer the rod and punishment, because force is used, and not love ; that the world could not go on thus, because now only ignorance, and not religion, is good ; that the Catholic religion pleased him more than the others, but that it had need of great formalities, which was not right, but very soon the world would see itself reformed, because it was impossible that such corruption should endure. He told me, too, that now, when the greatest ignorance flourishes which the world ever had, some glory in having the greatest knowledge there ever was, because they say they know what they do not understand, — which is, that God can be one and three, — and that these are impossibilities, ignorances, and most shocking blasphemies against the majesty of God. Besides this, he said that he liked women hugely, and that the Church committed a great sin in calling sin that which is according to nature."

After these charges, we hear no more of this latter-day Judas, Giovanni Mocenigo. Honest we can hardly deem him, for he confesses that he intended to betray Bruno long before he did betray him, and only delayed till he should gather sufficient damning evidence against him. And so we dismiss him to join the despicable crew of those who were traitors to their lords and benefactors.

The Inquisition examined four other witnesses. Two booksellers, Ciotto and Bertano, deposed that they had known Bruno at Frankfort-on-the-Main, whither they went to attend the famous bookfairs; that they had not heard him say aught which caused them to believe he was not a Catholic and a good Christian; but that he had the reputation of being a philosopher, who spent his time in writing and "in meditating new things." Andrea Morosini, a gentleman of noble birth, testified that during the recent months Bruno had been at his house, whither divers gentlemen and also prelates were wont to meet to discuss letters, and principally philosophy; but that he had never inferred from Bruno's remarks that he held opinions contrary to the faith. Finally, Fra Domenico da Nocera, of the Order of Preachers, deposed that "one day, near the feast of Pentecost, as I was coming out of the sacristy of the church of John and Paul, a layman, whom I did not know, bowed to me, and presently engaged in conversation. He said he was a friar of our province of Naples, a man of letters; Fra Giordano of Nola, his name. So we sought out a retired part of the aforesaid church. Then he told me how he had renounced the gown; of the many kingdoms he had traversed, and the royal courts, with his important exercises in letters; but that he had always lived as a Catholic. And I asking him what he was doing in Venice, and how he was living, he said that he had been in Venice but very few days, and was living comfortably; that he proposed to get tranquillity and write a book he had in his head, and to present it to his Holiness, for the quiet of his conscience and in order to be allowed to remain in Rome, and there devote himself to literary work, to show his ability, and perhaps to obtain a lectureship."

So far as we know, the Holy Office examined no other witnesses. That tribunal of the Inquisition at Venice was composed, in 1592, of the

Apostolic Nuncio, Monsignor Taberna; of the Patriarch, Monsignor Lorenzo Priuli; of the Father Inquisitor, Giovanni Gabriele da Saluzzo, a Dominican; and of three nobles appointed by the State, and called the *savii all' eresia* (sages or experts in heresy), who reported all proceedings to the Doge and Senate, and stopped the deliberations when they deemed them contrary to the laws and customs of the State, or to the secret instructions they had received. These three sages were, in that year, Luigi Foscari, Sebastian Barbarigo, and Tomaso Morosini.

Before this tribunal, which sat at the prison of the Inquisition, appeared the prisoner, Giordano Bruno, on Tuesday, May 26, 1592. He was a small, lean man, in aspect about forty years old, with a slight chestnut beard. On being bidden to speak, he began: —

" I will speak the truth. Several times I have been threatened with being brought to this Holy Office, and I have always held it as a jest, because I am ready to give an account of myself. While at Frankfort last year, I had two letters from Signor Giovanni Mocenigo, in which he invited me to come to Venice, as he wished me to teach him the art of memory and invention, promising to treat me well, and that I should be satisfied with him. And so I came, seven or eight months ago. I have taught him various terms pertaining to these two sciences; living at first outside of his house, and latterly in his own house. And, as it seemed to me that I had done and taught him as much as was necessary, and as was my duty in respect to the things he had sought me for, and deliberating, therefore, to return to Frankfort to publish certain of my works, I took leave of him last Thursday, so as to depart. He, hearing this, and doubting lest I wished to leave his house to teach other persons the very sciences I had taught him and others, rather than to go to Fraukfort, as I announced, was most urgent to detain me; but I none the less insisting on going, he began at first to complain that I had not taught him all I had agreed, and then to threaten me by saying that, if I would not remain of my own accord, he would find means to compel me. And the following night, which was Friday, seeing me firm in my resolution of going, and that I had put my things in order, and arranged to send them to Frankfort, he came, when I was in bed, with the pretext of wishing to speak to me; and after he had entered, there followed his servant Bortolo, with five or six others, who were, as I believed, gondoliers of the sort near by. And they made me get out of bed, and conducted me up to an attic, and locked me in there, Master Giovanni saying that, if I would remain and instruct him in the terms of memory and of geometry, as he had wished hitherto, he would set me at liberty; otherwise, something disagreeable would happen to me. And I replying all along that I thought I had taught him enough, and more than I was bound, and that I did not deserve to be treated in that fashion, he left me till the next day, when there came a captain, accompanied by certain men whom I did not know, and had them lead me down to a storeroom on the ground-floor of the house, where they left me till night. Then came another captain, with his assistants, and conducted me to the prison of this Holy Office, whither I believe I have been brought by the work of the aforesaid Ser Giovanni, who, indignant for the reason I have given, has, I think, made some accusation against me.

"My name is Giordano, of the Bruno family, of the city of Nola, twelve miles from Naples. I was born and brought up in that town; my profession has been, and is, that of letters and every science. My father's name was Giovanni, my mother's Fraulissa Savolina; he being a soldier by profession, who died at the same time with my mother. I am about forty-four years old, being born, according to what my people told me, in the year 1548. From my fourteenth year I was at Naples, to learn humanity, logic, and dialectics, and I used to attend the public lectures of a certain Sarnese; I heard logic privately from an Augustinian father, called Fra Theofilo da Vairano, who subsequently lectured on metaphysics at Rome. When I was fourteen or fifteen, I put on the habit of St. Dominick at the convent of St. Dominick at Naples. After the year of probation I was admitted to the profession, and then I was promoted to holy orders and to the priesthood in due time, and sang my first mass at Campagna, a town in the same kingdom. I lived there in a convent of the same order, called St. Bartholomew, and continued in this garb of St. Dominick, celebrating mass and the divine offices, and obedience to the superiors of the said order and of the priors of monasteries, till 1576, the year after the Jubilee. I was then at Rome, in the convent of the Minerva, under Master Sisto de Luca, procurator of the order, whither I had come because at Naples I had been brought to trial twice: the first time for having given away certain representations and images of the saints, and kept only a crucifix, wherefore I was charged with spurning the images of the saints; and, again, for saying to a novice, who was reading *The History of the Seven Joys* in verse, what business he had with such a book, — to throw it aside, and to read sooner some other work, like *The Lives of the Holy Fathers;* and this case was renewed against me at the time I went to Rome, together with other charges, which I do not know. On this account I left the order, and put off the gown.

"I went to Noli, in Genoese territory, and stayed there about four months, teaching small boys grammar, and reading lectures on the sphere [astronomy] to certain gentlemen; then I went away, first to Savona, where I tarried about a fortnight, and thence to Turin. Not finding entertainment there to my taste, I came to Venice by the Po, and lived a month and a half in the Frezzaria, in the lodging of a man employed at the Arsenal, whose name I do not know. Whilst I was here, I had printed this work [*On the Signs of the Times*], to make a little money for my support; I showed it first to Father Remigio de Fiorenza. Departing hence, I went to Padua, where I found some Dominican fathers, acquaintances of mine, who persuaded me to wear the habit again, even if I should not choose to return to the order; for it seemed to them more proper to wear that habit than not. With this view I went to Bergamo, and had made a garment of cheap white cloth, and over it I put the scapular, which I had kept when I left Rome. Thus attired I set out for Lyons; and at Chambèry, going to lodge with the order, and being very decently entertained, and talking about this with an Italian father who was there, he said to me, ' Be warned, for you will not meet with any sort of friendliness in these parts; and you will find less the farther you go.' So I set out for Geneva. There I lodged at the hostelry; and, a little after my arrival, the Marquis de Vico, a Neapolitan who was in that city, asked me who I was, and whether I had gone there to settle and to profess the religion of that place. I replied to him, after giving an account of myself and the reason why I had left the order, that I did not intend to profess that religion, because I did not know what it was; and that therefore I wished to abide there to live in liberty and to be safe. rather than for any other purpose. Being persuaded to put off that habit in any case, I took these clothes, and had a pair of hose made, and other things; and the marquis, with some other Italians, gave me a sword, hat, cloak, and other necessary articles, and, in order that I might support myself, they procured proof-reading for me. I kept to that work about two months, going, however, sometimes to preaching and sermons, whether of the Italians or of the French who lectured and preached there: among others, I heard more than once Nicolo Balbani, of Lucca, who read the Epistles of St. Paul, and preached on the Evangelists. But when I was told that I could not stay long in that place unless I should accept its religion, because I would have no employment from them, and finding, too, that I could not earn enough to live on, I went thence to Toulouse, where there is a famous university. Having become acquainted with some intelligent persons, I was asked to lecture on the sphere to divers students, which I did — with other lectures on philosophy — for perhaps six months. At this point, the post of 'ordinary' lecturer in philosophy, which is filled by competition, falling vacant, I took my doctor's degree, presented myself for the said competition, was admitted and approved, and lectured in that city two years continuously on the text of Aristotle's *De Anima* and other philosophical works. Then, on account of the Civil Wars, I quitted and went to Paris, where, in order to make myself known, and to give proof of myself, I undertook an 'extraordinary' lectureship, and read thirty lectures, choosing for subject Thirty Divine Attributes, taken from the first part of St. Thomas. Later, being requested to accept an 'ordinary' lectureship, I would not, because public lecturers in that city go generally to mass and the other divine offices, and I have always avoided this, knowing that I was excommunicated because I had quitted my order and habit; and although I had that 'ordinary' lectureship at Toulouse, I was not forced to go to mass, as I should have been at Paris. But conducting the 'extraordinary' there, I acquired such a name that the king, Henry III, sent for me, and wished to know whether my memory was natural or due to magic art. I satisfied him, both by what I said, and proved to him, that it was not by magic art, but by science. After this I published a work on the memory, under the title *De Umbris Idearum*, which I dedicated to his Majesty, — on which occasion he made me 'lecturer extraordinary,' with a pension; and I continued to read in that city perhaps five years, when, on account of the tumults which arose, I took my leave, and with letters from the king himself I went into England to reside with his ambassador, Michael de Castelnau. In his house I lived as a gentleman. I stayed in England two years and a half, and when the ambassador returned to France I accompanied him to Paris, where I remained another year. Having quitted Paris on account of the tumults, I betook myself to Germany, stopping first at Mayence, an archiepiscopal city, for twelve days. Finding neither here nor at Würzburg, a town a little way off, any entertainment, I went to Wittenberg, in Saxony, where I found two factions, — one of philosophers, who were Calvinists, the other of theologians, who were Lutherans. Among the latter was Alberigo

Gentile, whom I had known in England, a law-professor, who befriended me and introduced me to read lectures on the *Organon* of Aristotle; which I did, with other lectures in philosophy, for two years. At that time, the son of the old Duke having succeeded his father, who was a Lutheran, and the son being a Calvinist, he began to favor the party opposed to those who favored me; so I departed, and went to Prague, and stayed six months. Whilst there, I published a book on geometry, which I presented to the Emperor, from whom I had a gift of three hundred thalers. With this money, having quitted Prague, I spent a year at the Julian Academy in Brunswick; and the death of the Duke[1] happening at that time, I delivered an oration at his funeral, in competition with many others from the university, on which account his son and successor bestowed eighty crowns of those parts upon me; and I went away to Frankfort to publish two books, — one *De Minimo*, and the other *De Numero, Monade, et Figura*, etc. I stayed about six months at Frankfort, lodging in the convent of the Carmelites, — a place assigned to me by the publisher, who was obliged to provide me a lodging. And from Frankfort, having been invited, as I have said, by

[1] "Who was a heretic" is written on the margin of the original *procès-verbal*.

Ser Giovanni Mocenigo, I came to Venice seven or eight months ago, where what has since happened I have already related. I was going anew to Frankfort to print other works of mine, and one in particular on *The Seven Liberal Arts*, with the intention of taking these and some other of my published works which I approve — for some I do not approve — and of going to Rome to lay them at the feet of his Holiness, who, I have understood, loves the virtuous, and to put my case before him, with a view to obtain absolution from excesses, and permission to live in the clerical garb outside of the order. . . . I said I wish to present myself at the feet of his Holiness with some of my approved works, as I have some I do not approve, meaning by that that some of the works written by me and sent to the press I do not approve, because in them I have spoken and discussed too philosophically, unbecomingly, and not enough like a good Christian; and in particular I know that in some of these works I have taught and maintained philosophically things which ought to be attributed to the power, wisdom, and goodness of God according to the Christian faith; founding my doctrine on sense and reason, and not on faith. So much for them in general; concerning particulars, I refer to the writings, for I do not now recall a single article or particular doctrine I may have taught, but I will reply according as I shall be questioned and as I shall remember. . . .

"The subject of all my books, speaking broadly, is philosophy. In all of them I have always defined in the manner of philosophy and according to principles and natural light, not having most concern as to what, according to faith, ought to be believed; and I think there is nothing in them from which it can be judged that I professedly wish to impugn religion rather than to exalt philosophy, although I may have set forth many impious matters based on my natural light.

"I have taught nothing directly against Catholic Christian religion, although [I may have done so] indirectly; as was judged at Paris, where, however, I was allowed to hold certain disputes under the title of *One Hundred and Twenty Articles against the Peripatetics and Other Vulgar Philosophers* (printed with permission of the superiors); as it was permitted to treat them by the way of natural principles, without prejudice to the truth according to the light of faith, in which manner the books of Aristotle and Plato may be read and taught, which are in similar fashion, indirectly contrary to faith, — nay, much more so than the articles propounded and defended by me in the manner of philosophy: all these can be known from what is printed in my last Latin books from Frankfort, entitled *De Minimo, De Monade, de Immenso et Innumerabilibus*, and in part in *De Compositione Imaginum*. In these particularly you can see my intention and what I have held, which is, in a word, I believe in an infinite universe, — that is, the effect of infinite divine power; because I esteemed it unworthy of the divine goodness and power that, when it could produce besides this world another, and infinite others, it should produce a single finite world: so I have declared that there is an infinite number of particular worlds similar to this of the earth, which, with Pythagoras, I consider a star, like which is the moon, other planets, and other stars, which are infinite; and that all these bodies are worlds, without number, which make up the infinite university in infinite space, and we call this the infinite universe, in which are numberless worlds: so that there is a double infinitude, that of the greatness of the universe, and that of the multitude of the worlds, — by which indirectly it is meant to assail the truth according to faith.

"Moreover, in this universe I place a universal Providence, in virtue of which everything lives, vegetates, moves, and reaches its perfection; and I understand Providence in two ways: one in which it is present as the soul in all matter, and all in any part whatsoever, and this I call Nature, the shadow and footprint of the Deity; the other in the ineffable way with which God, by essence, presence, and power, is in all things and over all things, not as a part, but as Soul, in a manner indescribable. In the Deity I understand all the attributes to be one and the same substance, — just as theologians and the greatest philosophers hold; I perceive these attributes, power, wisdom, and goodness, or will, intelligence, and love, by means of which things have, first, being (by reason of the will), then, orderly and distinct being (by reason of the intelligence), and third, concord and symmetry (by reason of love); this I believe is in all and above all, as nothing is without participation in being, and being is not without its essence, just as nothing is beautiful without the presence of beauty; so nothing can be exempt from the divine presence. In this manner, by use of reason, and not by use of substantial [theological] truth, I discern distinctions in the Deity.

"Regarding the world as caused and produced, I meant that, as all being depends on the First Cause, I did not shrink from the term 'creation;' which I believe even Aristotle expressed, saying that God is, on whom the world and Nature are dependent; so that, according to the explanation of St. Thomas, be the world either eternal or temporal according to its nature, it is dependent on the First Cause, and nothing exists in it independently.

"Next, concerning that which belongs to faith — not speaking in the manner of philosophy — about the divine persons, that wisdom and that son of the mind, called by philosophers *intellect* and by theologians the *Word*, which we are to believe took upon itself human flesh, I, standing within the bounds of philosophy, have not understood it; but I have doubted, and with inconstant faith maintained, — not that I recall having shown a sign of it in writing or in speech, excepting as in other things indirectly one might gather from my belief and profession concerning those things which can be proved by the reason and deduced from natural light. And then concerning the divine spirit in a third person, I have been able to comprehend nothing in the way in which one ought to believe; but in the Pythagorean way, conformable to that way which Solomon points out, I have understood it to be the soul of the universe, or assistant in the universe, according to that saying in the *Wisdom of Solomon*, 'The Spirit of the Lord filleth the world; and that which containeth all things hath knowledge of the voice.'[1] This seems to me to agree to the Pythagorean doc-

[1] Chap. I, v, 7.

trine explained by Vergil in this passage of the *Æneid*: —

> 'Principio coelum ac terras camposque liquentes,
> Lucentemque globum Lunae Titaniaque astra,
> Spiritus intus alit, totamque infusa per artus
> Mens agitat molem.' [1]

"I teach in my philosophy that from this spirit, which is called the Life of the Universe, the life and soul of everything which has life and soul springs; that it is immortal, just as bodies, so far as concerns their substance, are all immortal, death being nothing else than division and coming together; this doctrine seems to be expressed in *Ecclesiastes*, where it says, 'There is no new thing under the sun. Is there anything whereof it may be said, See, this is new?' and so on."

Inquisitor. "Have you held, do you hold and believe, the Trinity, Father, Son, and Holy Ghost, one in essence, but distinct in person, as is taught and believed by the Catholic Church?"

Bruno. "Speaking as a Christian, and according to theology, and as every faithful Christian and Catholic ought to believe, I have indeed had doubts about the name 'person' as applied to the Son and the Holy Spirit; not understanding these two persons to be distinct from the Father, except as I have said above, speaking in the manner of

[1] Book VI, 724-27.

philosophy, and assigning the intelligence of the Father to the Son, and his love to the Holy Spirit, but without comprehending this word 'persons,' which in St. Augustine is declared to be not an ancient but a new word, and of his time: and I have held this opinion since I was eighteen years old till now, but in fact I have never denied, nor taught, nor written, but only doubted in my own mind, as I have said."

Inquisitor. "Have you believed, and do you believe, all that the Holy Mother Catholic Church teaches, believes, and holds about the First Person, and have you ever in any wise doubted concerning the First Person?"

Bruno. "I have believed and held undoubtingly all that every faithful Christian ought to believe and hold concerning the First Person. Regarding the Second Person, I declare that I have held it to be really one in essence with the First, and so the Third; because, being indivisible in essence, they cannot suffer inequality, for all the attributes which belong to the Father belong also to the Son and Holy Spirit: only I have doubted, as I said above, how this Second Person could become incarnate and could have suffered; nevertheless I have never denied nor taught that, and if I have said anything about this Second Person, I have said it in quoting the opinions of others, like Arius and

Sabellius and other followers of theirs. I will tell what I must have said, and which may have caused scandal and suspicion, as was set down in the first charges against me at Naples, to wit: I declared that the opinion of Arius seemed less pernicious than it was commonly esteemed and understood, because it is commonly understood that Arius meant to say that the Word is the first thing created by the Father; whereas I declared that Arius said that the Word was neither creator nor creature, but midway between creator and creature, — as the word is midway between the speaker and the thing spoken, — and therefore that the Word was the first-born before all creatures, not *by* which, but *through* which everything has been created, not *to* which but *through* which everything is referred and returns to the ultimate end, which is the Father. I exaggerated on this theme so that I was regarded with suspicion. I recall further to have said here in Venice that Arius did not intend to say that Christ, that is the Word, is a creature, but a mediator in the sense I have stated. I do not remember the precise place, whether at a druggist's or bookseller's, but I know I said this in one of these shops, arguing with certain priests who made a show of theology: I know not who they were, nor should I recognize them if I saw them. To make my statement more clear, I repeat that I have held there is one God, distinguished as Father, as Word, and as Love, which is the Divine Spirit, and that all these three are one God in essence; but I have not understood, and have doubted, how these three can get the name of persons, for it did not seem to me that this name of person was applicable to the Deity; and I supported myself in this by the words of St. Augustine, who says, '*Cum formidine proferimus hoc nomen personae, quando loquimur de divinis, et necessitate coacti utimur;*' besides which, in the Old and New Testaments I have not found nor read this expression nor this form of speech."

Inquisitor. "Having doubted the Incarnation of the Word, what has been your opinion about Christ?"

Bruno. "I have thought that the divinity of the Word was present in the humanity of Christ individually, and I have not been able to understand that it was a union like that of soul and body, but a presence of such a kind that we could truly say of this man that he was God, and of this divinity that it was man; because between substance infinite and divine and substance finite and human there is no proportion as between soul and body, or any other two things which can make up one existence; and I believe, therefore, that St. Augustine shrank from applying that word 'person' to this case: so that, in conclusion, I think, as regards my doubt of the Incarnation, I have wavered concerning its ineffable meaning, but not against the Holy Scripture, which says 'the Word is made flesh.'"

Inquisitor. "What opinion have you had concerning the miracles, acts, and death of Christ?"

Bruno. "I have held what the Holy Catholic Church holds, although I have said of the miracles that, while they are testimony of the divinity [of Christ], the evangelical law is, in my opinion, a stronger testimony, because the Lord said 'he shall do greater than these' miracles; and it occurred to me that whilst others, like the Apostles, wrought miracles, so that, in their external effect, they seemed like those wrought by him, Christ worked by his own virtue, and the Apostles by virtue of another's power. Therefore I have maintained that the miracles of Christ were divine, true, real, and not apparent; nor have I ever thought, said, nor believed the contrary.

"I have never spoken of the sacrifice of the mass, nor of transubstantiation, except in the way the Holy Church holds. I have believed, and do believe, that the transubstantiation of the bread and wine into the body and blood of Christ takes place really and in substance."

Inquisitor. "Did you ever say that Christ was

not God, but a good-for-nothing, and that, doing wretched works, he ought to have expected to be put to death, although he showed that he died unwillingly?"

Bruno. "I am astonished that this question is put to me, for I have never had such opinions, nor said such a thing, nor thought aught contrary to what I said just now about the person of Christ, which is that I believe what the Holy Mother Church believes. I know not how these things are imputed to me." At this he seemed much grieved.

Inquisitor. "In reasoning about the Incarnation of the Word, what have you held concerning the delivery of the said Word by the Virgin Mary?"

Bruno. "That it was conceived of the Holy Ghost, born of Mary as Virgin; and when any one shall find that I have said or maintained the contrary, I will submit myself to any punishment."

Inquisitor. "Do you know the import and effect of the sacrament of penance?"

Bruno. "I know that it is ordained to purge our sins; and never, never have I talked on this subject, but have always held that whosoever dies in mortal sin will be damned. It is about sixteen years since I presented myself to a confessor, except on two occasions: once at Toulouse, to a

Jesuit, and another time in Paris, to another Jesuit, whilst I was treating, through the Bishop of Bergamo, then nuncio at Paris, and through Don Bernardin de Mendoza, to reënter my order, with a view to confessing; and they said that, being an apostate, they could not absolve me, and that I could not go to the holy offices, wherefore I have abstained from the confessional and from going to mass. I have intended, however, to emerge some time from these censures, and to live like a Christian and a priest; and when I have sinned I have always asked pardon of God, and I would also willingly have confessed if I could, because I have firmly believed that impenitent sinners are damned."

Inquisitor. "You hold, therefore, that souls are immortal, and that they do not pass from one body into another, as we have information you have said?"

Bruno. "I have held, and hold, that souls are immortal, and that they are subsisting substances, that is rational souls, and that, speaking as a Catholic, they do not pass from one body into another, but go either to paradise or to purgatory, or to hell; but I have, to be sure, argued, following philosophical reasons, that as the soul subsists in the body, and is non-existent in the body [that is, not an integral part of it], it may, in the same way that it exists in one, exist in another, and pass from one to another; and if this be not true, it at least seems like the opinion of Pythagoras."

Inquisitor. "Have you busied yourself much in theological studies, and are you instructed in the Catholic resolutions?"

Bruno. "Not a great deal, having devoted myself to philosophy, which has been my profession."

Inquisitor. "Have you ever vituperated the theologians and their decisions, calling their doctrine vanity and other similar opprobrious names?"

Bruno. "Speaking of the theologians who interpret Holy Scripture, I have never spoken otherwise than well. I may have said something about some one in particular, and blamed him, — some Lutheran theologian, for instance, or other heretics, — but I have always esteemed the Catholic theologians, especially St. Thomas, whose works I have ever kept by me, read, and studied, and honored them, and I have them at present, and hold them very dear."

Inquisitor. "Which have you reckoned heretical theologians?"

Bruno. "All those who profess theology, but who do not agree with the Roman Church, I have esteemed heretics. I have read books by Melanchthon, Luther, Calvin, and by other heretics beyond the mountains, not to learn their doctrine nor to avail myself of it, for I deemed them more ignorant than myself, but I read them out of curiosity. I despise these heretics and their doctrines, because they do not merit the name of theologians, but of pedants; for the Catholic ecclesiastical doctors, on the contrary, I have the esteem I should."

Inquisitor. " How, then, have you dared to say that the Catholic faith is full of blasphemies, and without merit in God's sight ? "

Bruno. " Never have I said such a thing, neither in writing, nor in word, nor in thought."

Inquisitor. " What things are needful for salvation ? "

Bruno. " Faith, hope, and charity. Good works are also necessary ; or it will suffice not to do to others that which we do not wish to have done to us, and to live morally."

Inquisitor. " Have you ever denounced the Catholic religious orders, especially for having revenues ? "

Bruno. " I have never denounced one of them for any cause; on the contrary, I have found fault when the clergy, lacking income, are forced to beg ; and I was surprised, in France, when I saw certain priests going about the streets to beg, with open missals."

Inquisitor. " Did you ever say that the life of the clergy does not conform to that of the Apostles ? "

Bruno. " I have never said nor held such a thing ! " And as he said this he raised his hands, and looked about astonished. In answer to another question, he continued : " I have said that the Apostles achieved more by their preaching, good life, examples, and miracles than force can accomplish, which is used against those who refuse to be Catholics ; without condemning this method, I approve the other."

Inquisitor. " Have you ever said that the miracles wrought by Christ and the Apostles were apparent miracles, done by magic art, and not real ; and that you have enough spirit to work the same or greater, and wished finally to make the whole world run after you ? "

Bruno (lifting up both his hands). " What is this ? What man has invented this devilishness? I have never said such a thing, nor has it entered my imagination. O God, what is this ? I had rather be dead than that this should be proposed to me ! "

Inquisitor. " What opinion have you of the sin of the flesh, outside of the sacrament of matrimony ? "

Bruno. " I have spoken of this sometimes, saying, in general, that it was a lesser sin than the others, but that adultery was the chief of carnal sins, whereas the other was lighter, and almost venial. This, indeed, I have said, but I know and acknowledge to have spoken in error, because I remember what St. Paul says. However, I spoke thus through levity, being with others and discussing worldly topics. I have never said that the Church made a great mistake in constituting this a sin. . . .

"I hold it a pious and holy thing, as the Church ordains, to observe fasts and abstain from meat and prohibited food on the days she appoints, and that every faithful Catholic is bound to observe them; which I too would have done except for the reason given above; and God help me if I have ever eaten meat out of contempt [for the Church]. As for having listened to heretics preach, or lecture, or dispute, I did so several times from curiosity and to see their methods and eloquence, rather than from delight or enjoyment; indeed, after the reading or sermon, at the time when they distributed bread according to their form of communion, I went away about my business, and never partook of their bread nor observed their rites."

Inquisitor. "From your explanation of the Incarnation there follows another grave error, namely, that in Christ there was a human personality."

Bruno. "I recognize and concede that these and other improprieties may follow, and I have stated this opinion, not to defend, but only to explain it; and I confess my error such and so great as it is; and had I applied my mind to this adduced impropriety and to others deducible from it, I should not have reached these conclusions, because I may have erred in the premises, but certainly not in the conclusions."

Inquisitor. "Do you remember to have said that men are begotten of corruption, like the other animals, and that this has been since the Deluge down to the present?"

Bruno. "I believe this is the opinion of Lucretius. I have read it and heard it talked about, but I do not recall having referred to it as my opinion; nor have I ever believed it. When I reasoned about it, I did so referring it to Lucretius, Epicurus, and their similars, and it is not possible to deduce it from my philosophy, as will readily appear to any one who reads that."

Inquisitor. "Have you ever had any book of conjurations or of similar superstitious arts, or have you said you wished to devote yourself to the art of divination?"

Bruno. "As for books of conjurations, I have always despised them, never had them by me, nor attributed any efficacy to them. As for divination, particularly that relating to judicial astrology, I have said, and even proposed, to study it to see if there is any truth or conformity in it. I have communicated my purpose to several persons, remarking that, as I have examined all parts of philosophy, and inquired into all science except the judicial, when I had convenience and leisure I wish to have a look at that, which I have not done yet."

Inquisitor. "Have you said that the operations of the world are guided by Fate, denying the providence of God?"

Bruno. "This cannot be found either in my words or in my writings; on the contrary, you will find, in my books, that I set forth providence and free will. . . . I have praised many heretics and also heretic princes, but not as heretics, but only for the moral virtues they possessed. In particular, in my book *De la Causa, Principio et Uno*, I praise the Queen of England, and call her 'divine;' not as an attribute of religion, but as a certain epithet which the ancients used also to bestow on princes; and in England, where I then was and wrote that book, it is customary to give this title 'divine' to the Queen; and I was all the more persuaded to name her thus because she knew me, for I often went with the ambassador to court. I acknowledge to have erred in praising this lady, who is a heretic, and especially in attributing to her the epithet 'divine.'" . . .

Inquisitor. "Are the errors and heresies committed and confessed by you still embraced, or do you detest them?"

Bruno. "All the errors I have committed, down to this very day, pertaining to Catholic life and regular profession, and all the heresies I have held and the doubts I have had concerning the Catholic faith and the questions determined by the Holy Church, I now detest; and I abhor, and repent me of having done, held, said, believed, or doubted of anything that was not Catholic; and I pray this holy tribunal that, knowing my infirmities, it will please to accept me into the bosom of the Holy Church, providing me with remedies opportune for my safety and using me with mercy."

Bruno was then re-questioned concerning the reason why he broke away from his order. He repeated, in substance, the testimony already given, adding that his baptismal name was Philip.

Inquisitor. "Have you, in these parts, any enemy or other malevolent person, and who is he, and for what cause?"

Bruno. "I have no enemy in these parts, unless it be Ser Giovanni Mocenigo and his followers and servants, by whom I have been more grievously offended than by any other man living, because he has assassinated me in my life, in my honor, and in my goods, — having imprisoned me in his own house, confiscating all my writings, books, and other property; and he has done this, not only because he wished me to teach him all I knew, but also because he wished that I should not teach it to any one else; and he has always threatened my life and honor if I did not teach him what I knew."

Inquisitor. "Your apostacy of so many years renders you very suspicious to the Holy Faith, since you have so long spurned her censures, whence it may happen that you have held sinister opinions in other matters than those you have deposed; you may, therefore, and ought now to purify your conscience."

Bruno. "It seems to me that the articles I have confessed, and all that which I have expressed in my writings, show sufficiently the importance of my excess, and therefore I confess it, whatsoever may be its extent, and I acknowledge to have given grave cause for the suspicion of heresy. And I add to this that I have always had remorse in my conscience, and the purpose of reforming, although I was seeking to effect this in the easiest and surest way, still shrinking from going back to the straitness of regular obedience. . . . And I was at this very time putting in order certain writings to propitiate his Holiness, so that I might be allowed to live more independently than is possible as an ecclesiastic. . . .

"Beginning with my accuser, who I believe is Signor Giovanni Mocenigo, I think no one will be found who can say that I have taught false and heretical doctrine; and I have no suspicion that any one else can accuse me in matters of holy faith. It may be that I, during so long a course of time, may have erred and strayed from the Church in other matters than those I have exposed, and that I may be ensnared in other censures, but, though I have reflected much upon it, I have discovered nothing; and I now promptly confess my errors, and am here in the hands of your Excellencies to receive remedy, for my salvation. My force does not suffice to tell how great is my repentance for my misdeeds, nor to express it as I should wish." Having knelt down, he said: "I humbly ask pardon of God and your Excellencies for all the errors committed by me; and I am ready to suffer whatsoever by your prudence shall be determined and adjudged expedient for my soul. And I further supplicate that you rather give me a punishment which is excessive in gravity than make such a public demonstration as might bring some dishonor upon the holy habit of the order which I have worn; and if, through the mercy of God and of your Excellencies, my life shall be granted to me, I promise to make a notable reform in my life, and that I will atone for the scandal by other and as great edification."

Inquisitor. "Have you anything else to say for the present?"

Bruno. "I have nothing more to say."

II

This is the confession and apology of Giordano Bruno, taken from the minutes of the Inquisition of Venice, so far as I have been able to interpret the ungrammatical, ill-punctuated report of the secretary. The examinations were held on May 26 and 30, June 2, 3, 4, and July 30, 1592; and as there were, consequently, many repetitions of statement, I have condensed where it seemed advisable. From Bruno's lips we hear the explanation of his philosophical system, his doubts, his belief, and his recantation of any opinions which clashed with the dogmas of Catholicism. Was his recantation sincere? Before answering this question, let us glance at his opinions as he expressed them freely in his works; for upon Bruno's value as a thinker must finally rest the justification of our interest in him. True, the romance of his strange vagabond career and the pathos of his noble death will always excite interest in his personality; but the final question which mankind asks of prophet, philosopher, poet, preacher, or man of science is, "What can you tell us concerning our origin and our destiny?"

Be warned at the outset that Bruno furnished no complete, systematic reply to this question. He did not, like Spinoza, reduce his system to the precision of a geometrical text-book, all theorems and corollaries; nor, like Herbert Spencer, did he stow the universe away in a cabinet of pigeon-holes. He is often inconsistent, often contradicts himself. Perhaps his chief merit is that he stimulated thought on every subject he touched, and that he made sublime guesses which experiment, toiling patiently after him, has established as truths. Like all searchers for truth, his purpose was to discover the all-embracing Unity. Our reason shows us an unbridgeable chasm between matter and mind; the world of ideas and the outward world are in perpetual flux; nature is composed of innumerable separate objects, yet a superior unity pervades them. Life and death subsist antagonistically side by side: what is that, greater than both, which includes both? What is the permanence underlying this shifting, evanescent world? Conscience likewise reports the conflict between good and evil: what is the cause anterior to both? Many solutions have been offered; perhaps the commonest is that which, taught by the Manicheans and adopted by early Christians, announces that there are two principles in the universe, — one good, God, the other evil, Satan.

But insuperable difficulties accompany this view. If God be, as assumed, all-powerful, why does he not exterminate Satan; if he be just, why does he permit evil to exist at all?

Bruno, as we have seen in his deposition, proclaims that God is one and indivisible, the Soul of the universe; that his attributes are power, wisdom, and love; that he is in all things, yet above all things, not to be understood, ineffable, and whether personal or impersonal, man cannot say; that Nature is his footprint, God being the nature of Nature; that since every material atom is part of him, by virtue of his immanence in Nature, it is eternal, and so are human souls immortal, being emanations from his immortal spirit; but whether souls preserve their identity, or whether, like the atoms, they are forever re-composed into new forms, Bruno does not decide. This, speaking broadly, is pantheism; and pantheism is a system from which we are taught to recoil with almost as much horror as from atheism. "That is mere pantheism!" exclaimed John Sterling, aghast, at one of Carlyle's conclusions. "And suppose it were *pot*-theism? If the thing is true!" replied Carlyle, — a reply not to be taken for valid argument, perhaps, yet worthy of being pondered. As a pantheist, then, we must classify Bruno, — in that wide class which includes Spinoza, Goethe,

Shelley, and Emerson. " Within man is the soul of the whole," says Emerson; " the wise silence, the universal beauty, to which every part and particle is equally related, the eternal ONE. And this deep power in which we exist, and whose beatitude is all accessible to us, is not only self-sufficing and perfect in every hour, but the act of seeing and the thing seen, the seer and the spectacle, the subject and the object, are one." The Inquisition in 1600 would have burned Emerson for those two sentences.

Coming to details, we find that Bruno shakes himself free from the tyranny of Aristotle, — a mighty audacity, to measure which we must remember that upon Aristotle's arbitrary dicta the fathers and doctors of the Catholic Church had based their dogmas. Though a pagan, he had been for fifteen hundred years the logical pillar of Christendom, uncanonized, yet deserving canonization along with St. Thomas and St. Augustine. Bruno dared to attack the mighty despot in his very strongholds, the Sorbonne and Oxford, and, by so doing, helped to clear the road for subsequent explorers of philosophy and science. Equally courageous was his championship of the discoveries and theories of Copernicus. Bruno, we may safely say, was the first man who realized the full meaning of the Copernican system, — a meaning which even to-day the majority have not grasped. He saw that it was not merely a question as to whether the earth moves round the sun, or the sun moves round the earth; but that when Copernicus traced the courses of our solar system, and saw other and yet other systems beyond, he invalidated the strong presumption upon which dogmatic Christianity was reared. According to the old view, the earth was the centre of the universe, the especial gem of God's creation; as a final mark of his favor, God created man to rule the earth, and from among men he designated a few — his " chosen people " — who should enjoy everlasting bliss in heaven. But it follows from Copernicus's discoveries, that the earth is but one of a company of satellites which circle round the sun; that the sun itself is but one of innumerable other suns, each with its satellites; that there are probably countless inhabited orbs; that the scheme of salvation taught by the old theology is inadequate to the new conceptions we are bound to form of the majesty, justice, and omnipotence of the Supreme Ruler of an infinite universe. The God whom Bruno apprehended was not one who narrowed his interests to the concerns of a Syrian tribe, and of a sect of Christians on this little ball of earth, but one whose power is commensurate with infinitude, and who cherishes all creatures and all things in all worlds. Copernicus himself did not foresee the full significance of the discovery which dethroned the earth and man from their supposed preëminence in the universe; but Bruno caught its mighty import, and the labors of Kepler, Galileo, Newton, Herschel, and Darwin have corroborated him.

Inspired by this revelation, Bruno was the first to envisage religions as human growths, just as laws and customs are human growths, expressing the higher or lower needs and aspirations of the people and age in which they exist. His famous satire, *The Expulsion of the Beast Triumphant*,[1] has a far deeper purpose than to travesty classic mythology, or to ridicule the abuses of Romanists and Protestants, or to scoff at the exaggerated pretensions of the Pope. Under the form of an allegory, it is a prophecy of the ultimate passing away of all anthropomorphic religion. It shows how the god whom men have worshiped hitherto has been endowed by them with human passions and attributes, "writ large," to be sure, but still

---

[1] This, the most famous of Bruno's works, was until recently so rare that only two or three copies of it were known to exist. Hence numerous blunders and misconceptions by critics who wrote about it from hearsay. For a detailed analysis of "The Beast Triumphant" I may refer the reader to *The New World* for September, 1894. Lucian's satire, "Zeus in Heroics," may have given the hint to Bruno.

unworthy of being associated with that Soul of the World which is in all things, yet above all things. Everywhere he assails the doctrine that faith, without good works, can lead to salvation. He denounces celibacy, and other unnatural rules of the Catholic Church. He denounces still more vigorously the monstrous theory of original sin, according to which an assumedly just God punishes myriads of millions of human beings for the alleged trespass of two of their ancestors. Bruno also cites the discovery of new races in America as evidence that mankind are not all descended from Adam and Eve; whence he infers that, since the Mosaic cosmogony is too narrow to explain the creation and growth of mankind, the Hebrew scheme of vicarious punishment and vicarious redemption must be inadequate. He laughs at the idea of a "chosen people." Over and over again Bruno derides the assertion that, in order to be saved, we must despise our divinest guide, Reason, and be led blindly by Faith, reducing ourselves so far as we can to the level of donkeys. His satire, *La Cabala del Cavallo Pegaseo*, which supplements *The Beast Triumphant*, is a mock eulogy of this "holy asininity, holy ignorance, holy stupidity, and pious devotion, which alone can make souls so good that human genius and study cannot surpass them." "What avails, O truth-seeker," he exclaims in one of his finest sonnets, "your studying and wishing to know how Nature works, and whether the stars also are earth, fire, and sea? Holy donkeydom cares not for that, but with clasped hands wills to remain on its knees, awaiting from God its doom."

In a striking passage, Bruno explains that evil is relative. "Nothing is absolutely bad," he says; "because the viper is not deadly and poisonous to the viper, nor the lion to the lion, nor dragon to dragon, nor bear to bear; but each thing is bad in respect to some other, just as you, virtuous gods, are evil towards the vicious." Again he says, "Nobody is to-day the same as yesterday." The immanence of the universal soul in the animal world he illustrated thus: "With what understanding the ant gnaws her grain of wheat, lest it should sprout in her underground habitation! The fool says this is instinct, but we say it is a species of understanding."

These are some of Bruno's characteristic opinions. Their influence upon subsequent philosophers has been much discussed. His conception of the universe as an "animal" corresponds with Kepler's well-known view. Spinoza, the great pantheist of the following century, took from him the idea of an immanent God, and the distinction between *natura naturans* and *natura naturata*.

Schelling, who acknowledged Bruno as his master, found in him the principle of the indifference of contraries; Hegel, that of the absolute identity of subject and object, of the real and the ideal, of thought and things. La Croze discovers in Bruno the germs of most of Leibnitz's theories, beginning with the monad. Symonds declares that "he anticipated Descartes's position of the identity of mind and being. The modern theory of evolution was enunciated by him in pretty plain terms. He had grasped the physical law of the conservation of energy. He solved the problem of evil by defining it to be a relative condition of imperfect energy. . . . We have indeed reason to marvel how many of Bruno's intuitions have formed the stuff of later, more elaborated systems, and still remain the best which these contain. We have reason to wonder how many of his divinations have worked themselves into the common fund of modern beliefs, and have become philosophical truisms."[1] Hallam, who strangely undervalued Bruno, states that he understood the principle of compound forces. After making due allowance for the common tendency to read back into men's opinions interpretations they never dreamed of, we shall find that much solid sub-

---

[1] From J. A. Symonds's *Renaissance in Italy: The Catholic Reaction*, chap. ix.

stance still remains to Bruno's credit. He is, above all, suggestive.

III

We come now to that perplexing question, "Why did he recant? How could he, who was so evidently a freethinker and a rationalist, honestly affirm his belief in the Roman Catholic dogmas?" His confession seems to be straightforward and candid: had he wished to propitiate the Inquisitors, he needed only not to mention his philosophical doubts about the Incarnation and the Trinity; he needed only to admit that there were in his writings errors which he no longer approved, and to throw himself on the mercy of his tribunal. What, then, was the motive? Was it physical fear? Did life and liberty seem too tempting to him who loved both so intensely; preferable to death, no matter how great the sacrifice of honor? Did he simply perjure himself? Or was he suddenly overcome by a doubt that his opinions might be, after all, wrong, and that the Church might be right? He testified, and others testified, that before he had any thought of being brought to trial he had determined to make his peace with the Pope, and to obtain leave, if he could, to pass the remainder of his life in philosophical tranquillity. Did the early religious associations and prejudices, which he supposed had long ago ceased to influence him, unexpectedly spring up, to reassert a temporary tyranny over his reason? Many men not in jeopardy of their lives have had this experience of the tenacious vitality of the doctrines taught to them before they could reason. Did it seem to him a huge Aristophanic joke that a church which then had but little real faith and less true religion in it should call any one to account for any opinions, and that therefore the lips might well enough accept her dogmas without binding the heart to them? Many men, who believed themselves sincere, have subscribed in a "non-natural sense" to the Thirty-nine Articles of Anglicanism; did Bruno subscribe to the Catholic Articles under a similar mental reservation? Or, believing, as he did, that every religion contains fragments of the truth, could he not honestly say he believed in Catholicism, at the same time holding that her symbols had a deeper significance than her theologians perceived, and that the truth he apprehended was immeasurably wider?—just as a mathematician might subscribe to the multiplication table, knowing that it is not the final bound of mathematical truth, but only the first step towards higher and unlimited investigations.

Throughout his examination Bruno was careful to make the distinction between the province of faith and the province of speculation. "Speaking after the manner of philosophy," he confessed that he had reached conclusions which, "speaking as a Catholic," he ought not to believe. This distinction, which we now think uncandid and casuistical, was nevertheless admitted in his time. All through that century, men had argued "philosophically" about the immortality of the soul; but "theologically" such an argument was impossible, because the Church pronounced the immortality of the soul to be an indisputable fact. But, we ask, can a man honestly hold two antagonistic, mutually destroying beliefs; saying, for instance, that his reason has disproved the Incarnation, but that his faith accepts that doctrine? Or was Bruno unaware of his contradictions? Of how many of your opinions concerning the ultimate mysteries of life do you, reader, feel so sure that, were you suddenly seized, imprisoned, brought face to face with a pitiless tribunal, and confronted by torture and burning, you — one man against the world — would boldly, without hesitation, publish and maintain them? Galileo, one of mankind's noblest, could not endure this ordeal, although the evidence of his senses and the testimony of his reason contradicted the denial which pain and dread wrung from him. Savonarola, another great spirit, flinched likewise. These are points we are bound to consider before we pronounce Bruno a hypocrite or a coward.

The last news we have of him in Venice is when, "having been bidden several times," he rose from his knees, after confessing his penitence, on that 30th of July, 1592. The authorities of the Inquisition at Rome immediately opened negotiations for his extradition. The Doge and Senate demurred; they hesitated before establishing the precedent whereby Rome could reach over and punish Venetian culprits. Time was, indeed, when Venice allowed no one, though he were the Pope, to meddle in her administration; but, alas! the lion had died out in Venetian souls. Finally, "wishing to give satisfaction to his Holiness," Doge and Senators consented to deliver Bruno up; the Pope expressed his gratification, and said that he would never force upon the Republic "bones hard to gnaw." So Bruno was taken to Rome. In the "list of the prisoners of the Holy Office, made Monday, April 5, 1599," we find that he was imprisoned on February 27, 1593. What happened during almost seven years we can only surmise. We may be sure the Inquisitors searched his books for further heretical doctrine. We hear that they visited him in his cell from time to time, and exhorted him to recant, but that he replied that he had nothing to abjure, and that they had misinterpreted him. A memorial which he addressed to them they did not read. Growing weary of their efforts to save his soul, they would temporize no more; on a given day he must retract, or be handed over to the secular arm. That day came: Giordano Bruno stood firm, though he knew the penalty was death.

We cannot tell when he first resolved to dare and suffer all. Some time during those seven years of solitude and torment, he awoke to the great fact that

> " 'T is man's perdition to be safe,
> When for the truth he ought to die."

Mere existence he could purchase with the base coin of cowardice or casuistry; but that would be, not life, but a living shame, and he refused. Who can tell how hard instinct pleaded, — how the thoughts of freedom, how the longings for companions, how the recollections of that beautiful Neapolitan home which he loved and wished to revisit, how the desire to explore yet more freely the beauties and the mysteries of the divine universe, came to him with reasons and excuses to tempt him from his resolution? But conscience supported him. He took Truth by the hand, turned his back on the world and its joy and sunshine, and followed whither she led into the silent, sunless unknown. Let us dismiss the theory that he was impelled by the desire to escape in this way from an imprisonment which threatened to be perpetual; let us dismiss, and contemptuously dismiss, the insinuation of an English writer, that Bruno's purpose was, by a theatrical death, to startle the world which had begun to forget him in his confinement. To impute a low motive to a noble deed is surely as base as to extenuate a crime. Bruno had no sentimental respect for martyrs; but on the day when he resolved to die for his convictions, he proved his kinship with the noblest martyrs and heroes of the race.

On February 8, 1600, he was brought before Cardinal Mandruzzi, the Supreme Inquisitor. He was formally degraded from his order, sentence of death was pronounced against him, and he was given up to the secular authorities. During the reading, he remained tranquil, thoughtful. When the Inquisitor ceased, he uttered those memorable words, which still, judging from the recent alarm in the Vatican, resound ominously in the ears of the Romish hierarchy: "Peradventure you pronounce this sentence against me with greater fear than I receive it." After nine days had been allowed for his recantation, he was led forth, on February 17, to the Campo di Fiora, — once an amphitheatre, built by Pompey, and now a vegetable market. When he had been bound to the stake, he protested, according to one witness, that he died willingly, and that his soul would mount with the smoke into paradise. Another account says that he was gagged, to prevent his uttering blasphemies. As the flames leaped up, a crucifix was held before him, but he turned his head away. He uttered no scream, nor sigh, nor murmur, as Hus and Servetus had done; even that last mortal agony of the flesh could not overcome his spirit. And when nothing remained of his body but ashes, these were gathered up and tossed to the winds.

Berti, to whose indefatigable and enlightened researches, extending over forty years, we owe our knowledge of Bruno's career,[1] says justly that Bruno bequeathed to his countrymen the example of an Italian dying for an ideal, — a rare example in the sixteenth century, but emulated by thousands of Italians in the nineteenth. To us and to all men his death brings not only that lesson, but it also teaches that no tribunal, whether religious or political, has a right to coerce the conscience and inmost thoughts of any human being. Let a man's deeds, so far as they affect the community, be amenable to its laws, but his opinions should

---

[1] See Berti's work, *Giordano Bruno da Nola; Sua Vita e Sua Dottrina*, 1889. This excellent biography deserves to be translated into English.

be free and inviolable. We can grant that the Torquemadas and Calvins and Loyolas were sincere, and that, from their point of view, they were justified in persecuting men who differed from them in religion; for the heretic, they believed, was Satan's emissary, and deserved no more mercy than a fever-infected rag; but history admonishes us that their point of view was not only cruel, but wrong. No man, no church, is infallible: therefore it may turn out that the opinions which the orthodoxy of yesterday deemed pernicious have infused new blood into the orthodoxy of to-day. Bruno declared that the universe is infinite and its worlds are innumerable; the Roman Inquisition, in its ignorance, knew better. Galileo declared that the earth moves round the sun; the Inquisition, in its ignorance, said, No. It burned Bruno, it harried Galileo; yet, after three centuries, which do we believe? And if the Roman Church was fallible in matters susceptible of easy proof, shall we believe that it, or any other church, is infallible in matters immeasurably deeper and beyond the scope of finite demonstration? Cardinal Bellarmine, an upright man, and perhaps the ablest Jesuit of any age, was the foremost Inquisitor in bringing Bruno to the stake, and in menacing Galileo with the rack; but should a schoolboy of ten now uphold Bellarmine's theory of the solar system, he would be sent into the corner with a fool's-cap on his head.

Strange is it that mankind, who have the most urgent need for truth, should have been in all ages so hostile to receiving it. Starving men do not kill their rescuers who bring them bread; whereas history is little more than the chronicle of the persecution and slaughter of those who have brought food for the soul. Doubtless the first savage who suggested that reindeer-meat would taste better cooked than raw was slain by his companions as a dangerous innovator. Ever since that time, the messengers of truth have been stoned, and burned, and ganched, and crucified; yet their message has been delivered, and has at last prevailed. This is, indeed, the best encouragement we derive from history, and the fairest presage of the perfectibility of mankind.

The study of the works of Giordano Bruno, which has been revived and extended during this century, is one evidence of a more general toleration, and of a healthy desire to know the opinions of all kinds of thinkers. One reason why Bruno has attracted modern investigators is because so many of his doctrines are in tune with recent metaphysical and scientific theories; and it seems probable that, for a while at least, the interest awakened in him will increase rather than diminish, until, after the republication and examination of all his writings, a just estimate of his speculations shall have been made. Much will undoubtedly have to be thrown out as obsolete or fanciful; much as flippant and inconsistent; much as vitiated by the cumbrous methods of scholasticism, and the tedious fashion of expounding philosophy by means of allegory and satire. But, after all the chaff has been sifted and all the excrescences have been lopped off, something precious will remain.

The very diversity of opinions about the upshot and value of his teaching insures for him the attention of scholars for some time to come. Those thinkers who can be quickly classified and easily understood are as quickly forgotten; only those who elude classification, and constantly surprise us by turning a new facet towards us, and provoke debate, are sure of a longer consideration. And see how conflicting are the verdicts passed upon Bruno. Sir Philip Sidney and that fine group of men who just preceded the Shakespearean company were his friends, and listened eagerly to his speculations. Hegel says: " His inconstancy has no other motive than his great-hearted enthusiasm. The vulgar, the little, the finite, satisfied him not; he soared to the sublime idea of the Universal Substance." The French *philosophes* of the eighteenth century debated whether he were an atheist; the critics of the nineteenth century declare him to be a pantheist. Hallam thought that, at the most, he was but a "meteor of philosophy." Berti ranks him above all the Italian philosophers of his epoch, and above all who have since lived in Italy except Rosmini, and perhaps Gioberti. Some have called him a charlatan; some, a prophet. Finally, Leo XIII, in an allocution which was read from every Romish pulpit in Christendom, asserted that "his writings prove him an adept in pantheism and in shameful materialism, imbued with coarse errors, and often inconsistent with himself;" and that "his talents were to feign, to lie, to be devoted wholly to himself, not to bear contradiction, to be of a base mind and wicked heart." As we read these sentences of Leo XIII, and his further denunciation of those who, like Bruno, ally themselves to the Devil by using their reason, we reflect that, were popes as powerful now as they were three centuries ago, they would have found reason enough to burn Mill and Darwin, and many another modern benefactor.

Bruno's character, like his philosophy, offers so many points for dispute that it cannot soon cease to interest men. He is so human — neither demigod nor demon, but a creature of perplexities and contradictions — that he is far more fascinating than those men of a single faculty, those monotones whom we soon estimate and tire of. His vitality, his daring, his surprises, stimulate us. In an age when the growing bulk of rationalism casts a pessimistic shadow over so many hopes, it is encouraging to know that the rationalist Bruno saw no reason for despair; and when some persons are seriously asking whether life be worth living, it is inspiring to point to a man to whom the boon of life was so precious and its delights seemed so inexhaustible. At any period, when many minds, after exploring all the avenues of science, report that they perceive only dead matter everywhere, it must help some of them to learn that Bruno beheld throughout the whole creation and in every creature the presence of an infinite Unity, of a Soul of the World, whose attributes are power, wisdom, and love. He was indeed "a God-intoxicated man." Aristotle, Ptolemy, and Aquinas spun their cobwebs round the border of the narrow circle in which, they asserted, all truth, mundane and celestial, was comprehended; Bruno's restless spirit broke through the cobwebs, and discovered limitless spaces, innumerable worlds, beyond. To his enraptured eyes, all things were parts of the One, the Ineffable. "The Inquisition and the stake," says Mr. Symonds, "put an end abruptly to his dream. But the dream was so golden, so divine, that it was worth the pangs of martyrdom. Can we say the same for Hegel's system, or for Schopenhauer's, or for the encyclopædic ingenuity of Herbert Spencer?" By his death Bruno did not prove that his convictions are true, but he proved beyond peradventure that he was a true man; and by such from the beginning has human nature been raised towards that ideal nature which we believe divine.

BRYANT

THERE are many good reasons why we should celebrate the one hundredth birthday of William Cullen Bryant.[1] Not the least of them is this, that in bringing him our tribute we also commemorate the birthday of American poetry. He was our earliest poet, and "Thanatopsis" our earliest poem. Through him, therefore, we make festival to the Muse who has taught many since him to sing.

Older than Bryant were three single-poem men, — Francis Scott Key, Joseph Hopkinson, and John Howard Payne; yet, so far as I can learn, their three poems were written later than "Thanatopsis," and, after all, neither "The Star Spangled Banner," nor "Hail Columbia," nor "Home, Sweet Home," would rank high as poetry. Likewise, though Fitz-Greene Halleck was older than Bryant by four years, and once enjoyed a considerable vogue, his verse is now obsolescent, if not obsolete. In the anthologies — those presses of faded poetical flowers — you will still find some

[1] First printed in *The Review of Reviews*, New York, October, 1894.

of his pieces; but which of us now regards "Marco Bozzaris" as the "finest martial poem in the language"?

Bryant's priority among his immediate contemporaries is thus clearly established; furthermore, a considerable interval separated him from that group of American poets who rose to eminence in the two decades before the civil war. Bryant was born in 1794, Emerson in 1803, Longfellow and Whittier in 1807, Holmes and Poe in 1809, Lowell and Whitman in 1819. An almost unexampled precocity also set Bryant's pioneership beyond dispute.

But when we call Bryant the earliest American poet, and "Thanatopsis" the earliest American poem, we must not suppose that both had not had many ineffectual predecessors. Versifiers, like milliners, flourish from age to age, and their works are forgotten in favor of a later fashion. Who the forgotten predecessors of Bryant were, he himself will tell us. Being asked in February, 1818, to write an article on American poetry for the *North American Review* he replied: —

"Most of the American poets of much note, I believe, I have read, — Dwight, Barlow, Trumbull, Humphreys, Honeywood, Clifton, Paine. The works of Hopkins I have never met with. I have seen Philip Freneau's writings, and some things by

Francis Hopkinson. There was a Dr. Ladd, if I am not mistaken in the name, of Rhode Island, who, it seems, was much celebrated in his time for his poetical talent, of whom I have seen hardly anything; and another, Dr. Church, a Tory at the beginning of the Revolution, who was compelled to leave the country, and some of whose satirical verses which I have heard recited possess considerable merit as specimens of forcible and glowing invective. I have read most of Mrs. Morton's poems, and turned over a volume of stale and senseless rhymes by Mrs. Warren. Before the time of these writers, some of whom are still alive, and the rest belong to the generation which has just passed away, I imagine that we could hardly be said to have any poetry of our own; and indeed it seems to me that American poetry, such as it is, may justly enough be said to have had its rise with that knot of Connecticut poets, Trumbull and others, most of whose works appeared about the time of the Revolution."[1]

Bryant's list contains the name of not one poet whose works are read to-day. All these volumes belong to *fossil literature*, — literature, that is, which may be dug up and studied for the light it may throw on the customs of a time, or its intel-

---

[1] *A Biography of William Cullen Bryant*, by Parke Godwin, i, 154.

lectual development, but which, so far as its own vitality is concerned, has passed away beyond hope of resuscitation. The historical student of American poetry may read Barlow's "Columbiad" as a matter of duty; but those of us to whom poetry is the breath of life will not seek it in that literary graveyard. Reverently, rather, will we read the titles on the tombstones and pass on.

Almost coeval with American independence itself was the notion that there ought to be an independent American literature. The Revolution had resulted in the formation of a republic new in pattern, in opportunities, in ideals; a republic which, having broken forever with the political system of Britain, would gladly have been freed from all obligations — including intellectual and æsthetic obligations — to her. We hardly realize how acute was the sensitiveness of our great-grandfathers on this point. The satisfaction they took in recalling the victories of Bennington and Yorktown vanished when they were reminded — and there was always some candid foreigner at hand to remind them — that a nation's real greatness is measured, not by the size of its crops, nor by its millions of square miles of surface, nor by the rapidity with which its population doubles, nor even by its ability to whip King George the Third's armies, but by its contributions to philosophy, to literature, to art, to religion. "What have you to show in *these* lines?" we imagine the candid foreigner to have been perpetually asking; and the patriotic American to have winced, as he had to reply, "Nothing;" unless, indeed, he happened to have Thomas Jefferson's philosophical poise. To the slur of Abbé Raynal, that "America had not produced a single man of genius," Jefferson replied: "When we shall have existed as a people as long as the Greeks did before they produced a Homer, the Romans a Virgil, the French a Racine and Voltaire, the English a Shakespeare and Milton, should this reproach be still true, we will inquire from what unfriendly causes it has proceeded that the other countries of Europe and other quarters of the earth shall not have inscribed any name of ours on the roll of poets."

Very few Americans, however, could bear with Jeffersonian equanimity the imputation of inferiority. All were well aware that they had just achieved a revolution without parallel in history; they were honestly proud of it; and they could not help feeling touchy when their critics, ignoring this stupendous achievement, censured them for failure in fields they had never entered. A few, like Jefferson, would respond, "Give us time;" the majority either masked their irritation under pretended contempt for the opinion of foreigners, or silently admitted the impeachment. There grew up, on the one hand, "spread-eagleism," — brag over our material and political bigness, — and, on the other, an impatient desire to produce masterpieces which should not fear comparison with the best the world could show. The Hebrew patriarchs, whose faith Jehovah tested by denying them children till the old age of their wives, were not less troubled at the postponement of their dearest wishes than were those eager watchers for the advent of American genius. Long before Bryant's little volume was published, in 1821, those watchers had begun to speculate as to the sort of work in which that genius would manifest itself, and then was conjured up that bogy, "The American Spirit," which has flitted up and down through our college lecture-rooms and fluttered the minds of immature critics ever since. It was generally agreed that the question to be asked about each new book should be, " Has it The American Spirit?" and not, "Is it excellent?" Nobody knew how to define that spirit, but everybody had a teasing conviction that, unless it were conspicuous, the offspring of American genius could not prove their legitimacy. Foreigners, especially the English, encouraged this conviction. They expected something strange and uncouth; they would accept nothing else as genuine. Hence, years afterward, when Whitman, with cowboy gait, came swaggering up Parnassus, shouting nicknames at the Muses and ready to slap Apollo on the back, our perspicacious English cousins exclaimed, "There! there! that's American! At last we've found a poet with The American Spirit!" For quite other reasons Whitman deserves serious attention; not for those extravagances which he deluded himself and his unrestrained admirers into thinking were most precious manifestations of The American Spirit. This bogy has now been pretty thoroughly exorcised, its followers being chiefly the writers of bad grammar, bad spelling, and slang, — which pass for dialect stories, — and an occasional student of literature, who finds very little of the American product that could not have been produced elsewhere. We may dismiss The American Spirit, bidding it seek its spectral companion, The Great American Novel, but we must remember that, even before Bryant began to write, it was worrying the minds of our literary folk.

Bryant himself must have been subjected, consciously or unconsciously, to the influences we have surveyed, — for who can escape breathing the common atmosphere? But he had within him that which is more potent than any external mould, and is the one trait hereditary in genius of every kind, — he had sincerity. What he saw, he saw with his own eyes; what he spake, he spake with his own lips; and inevitably it followed that men proclaimed him original. His secret, his method, were no more than this. "I saw some lines by you to the skylark," he writes to his brother in 1838. "Did you ever see such a bird? Let me counsel you to draw your images, in describing Nature, from what you observe around you, unless you are professedly composing a description of some foreign country, when, of course, you will learn what you can from books. The skylark is an English bird, and an American who has never visited Europe has no right to be in raptures about it." That last sentence explains Bryant; it is worth a hundred essays on The American Spirit; it should be the warning of every writer. The raptures of Americans over English skylarks they had never seen were then, and have always been, the bane of our literature. Eighty years ago the lowlands at the foot of our Helicon had been turned into a slough by the tears of rhymsters who did not feel the griefs they sang of, and the woods howled with sighs which caused no pang to the sighers. Bryant, by merely being natural and sincere, was instantly recognized as belonging to that lineage every one of whose children is a king.

The story of his entry into literature, though

well known, cannot be too often told. Born at Cummington, a little village on the Hampshire hills, Massachusetts, November 3, 1794, his father was a genial, fairly cultivated country doctor; his mother, Sarah Snell, an indefatigable housewife, with Yankee common-sense and deep-grained Puritan principles. William Cullen, the second of several children, was precocious; both parents encouraged his aptitude for verse-making, and a satire which he wrote in 1807 on Jefferson and the Embargo his father was proud to have printed in Boston. In 1810 young Bryant entered the sophomore class of Williams College, and spent a year there. He hoped to pass from Williams to Yale, where he looked for more advanced instruction, but his father's means did not permit, and the son, instead of finishing his course at Williams, went into a country lawyer's office and fitted himself for the bar. Just at the moment of indecision, in the autumn of 1811, Bryant wrote "Thanatopsis." Contrary to his custom, he did not show it to his father, but laid it away with other papers in a drawer. Six years later Dr. Bryant, whose duties as a member of the Massachusetts legislature took him often to Boston, and whose bright parts and liberal views made him welcome in the foremost circles there, was asked by his friends, who edited the *North American*

Review, for some contribution. On returning to Cummington, he happened to find his son's sequestered papers, and, choosing "Thanatopsis" — of which, the original being covered with many corrections, he made a copy — and "The Waterfowl," he sent them off to Boston, and they appeared in the *Review* for September, 1817. The young poet, having meanwhile completed his legal studies, was practicing law at Great Barrington, unconscious of the fame about to descend upon him. Owing to the handwriting of the copy of the poems sent to the *Review*, however, Dr. Bryant had for a moment the credit of being the author of "Thanatopsis."

After duly allowing for the common tendency to make fame retroactive, we cannot doubt that "Thanatopsis" secured immediate and, relatively, immense recognition. The best judges agreed that at last a bit of genuine American literature was before them; the uncritical but appreciative, from ministers to school children, read, learned, admired, and quoted the grave, sonorous lines.

Thanatopsis, — a Vision of Death! A strange corner-stone for the poetic literature of the nation which had only recently sprung into life, — a nation conscious as no other had been of its exuberant vitality, of its boundless material resources, of its expansiveness and invincible will. Yet neither the glory achieved nor the ambition cherished fired the imagination of the youthful poet. He looked upon the earth, and saw it but a vast grave; he looked upon men and beheld, not their high ambitions nor the great deeds which blazon human story, but their transcience, their mortality. Nothing in life could so awe him as the majestic mystery of death.

The mood, I believe, is not rare among sensitive and thoughtful youths, who, just as their faculties have ripened sufficiently to enable them to feel a little of the unspeakable delight of living, are staggered at realizing for the first time that death is inevitable, and that the days of the longest life are few. That this terrific discovery should kindle thoughts full of sublimity need not surprise us; but we may well be astonished that Bryant at seventeen should have had power to express them in a poem which is neither morbid nor religiously commonplace.

In 1821 Bryant received what was then the blue ribbon of recognition in being asked to deliver a poem before the Harvard Chapter of the Phi Beta Kappa Society. He wrote "The Ages," read it in Cambridge and printed it, together with "Thanatopsis" and a few other pieces, in a little volume. The previous conviction was confirmed; every one spoke of Bryant as *the* American poet. Even the professional critics — those sapient fellows whose obtuseness is the wonder of posterity, the clique which pooh-poohed Keats, and ha-hahed Wordsworth, and bear-baited Carlyle — made in Bryant's case no mistake. Although one of them, indeed, declared that there was "no more poetry in Bryant's poems than in the Sermon on the Mount," yet the opinions were generally laudatory, and the critics were quick in defining the qualities of the new poet. They found in him something of Cowper and something of Wordsworth, but the resemblances did not imply imitation; Bryant might speak their language, but it was his also. No one questioned the genuineness of his inspiration, and not for a quarter of a century after the publication of "Thanatopsis," that is, not until the early forties, — when Longfellow, Whittier, Poe, and Emerson began to have a public for their poetry, — did any one question Bryant's primacy. He had been so long the *only* American poet that it was naturally assumed that he would always be the *best*. He had redeemed America from the reproach of barrenness in poetry, as Irving and Cooper redeemed its prose, and Americans could feel toward no others as they felt toward him.

A hundred years have elapsed since his birth; three generations have known his works: what is Bryant to us, who are posterity to him? Is he, like Cimabue in painting, a mere name to date from, — a pioneer whom we respect, — and nothing more? Far from it. Bryant's poetry is not only chronologically but absolutely interesting: it lives to-day, and the qualities which have vitalized it for three quarters of a century show no signs of decay. It would be incorrect, of course, to assert that Bryant holds relatively so high a place in our literature as he held fifty years ago; his estate then was the first poetic clearing in the wilderness; its boundaries are still the same; but subsequent poets have made other clearings all round his, and brought different prospects into view and different talents under cultivation.

Let us look briefly at Bryant's domain. Intimate and faithful portrayal of Nature is the product which first draws our attention; next we perceive that the observer who makes the picture is a sober moralist. He delights in Nature for her own sake, for her beauty and variety; and then she suggests to him some rule of conduct, some parallel between her laws and the laws of human life, by which he is comforted and uplifted. Bryant, I have said elsewhere, interprets Nature morally, Emerson spiritually, and Shelley emotionally. We need not stop to inquire which of these methods of interpretation is the highest. Suffice it for us to realize that all of them are valuable, and that the poet who succeeds in identifying himself in a marked degree with any one of them will not soon be forgotten.

That Wordsworth preceded Bryant in the moral interpretation of Nature detracts nothing from Bryant's merit. The latest prophet is no less original than the earliest; for originality lies in being a prophet at all. Young Bryant, wandering over the bleak Hampshire hills or in the woods or along the brawling streams, had original impressions, which he trustingly recorded; and to-day, if you go to Cummington, you will marvel at the fidelity of his record. But his poetry is true not only there; it is true in every region where Nature has similar aspects; symbolically, it is true everywhere.

There being no doubt as to the veracity of his pictures, what shall we say of that other quality, the moral tone which pervades them? That, too, is of a kind men will not soon outgrow. It inculcates courage, patience, fortitude, trust; it springs from the optimism of one who believes in the ultimate triumph of good, not because he can prove it, but because his whole being revolts at the thought of evil triumphant. He has the stoic's dread of flinching before any shock of misfortune, the Christian's dread of the taint of sin. Here are two ideals, each the complement of the other, which the world cannot outgrow, and the poet who

— pondering on a fringed gentian or the flight of a waterfowl, or on a rivulet bickering among its grasses — found new incitements to courage and virtue, thereby associated himself with the eternal. To interpret nature morally in this fashion, which is Bryant's fashion, is to rise far above the level of the common didacticism of our pulpits. Professional moralists go to nature for figures of speech to furnish forth their sermons and religious verse, as they go to their kitchen garden for vegetables; but they do not enter Bryant's world.

Moreover, in painting the scenery of the Hampshire hills, and in saturating his descriptions with the moral tonic I have spoken of, Bryant became the representative of a phase of New England life which has had an incalculable influence on the development of this nation. The mitigated Spartanism amid which his youth was passed bred those colonists who carried New England standards with them to the shores of the Pacific. A Puritan by derivation and environment, Bryant was by training and conviction a Unitarian, — a combination which made him in a sense the exemplar both of the austerity which had characterized New England ideals in the past, and of the liberalism which during this century has nowhere found more strenuous supporters than in New England.

On many positive grounds, therefore, Bryant's

title to fame rests; he was one of Nature's men, he shed moral health, he uttered the ideals of a great race in a transitional epoch. His temperament, in making his poetic product small, gave him yet another hostage against oblivion. The poet who, having so many claims to the consideration of posterity, can also plead brevity, need not worry himself about what is called literary immortality. Bryant's typical and best work is comprised in a dozen poems, the longest not exceeding 140 lines. Read "Thanatopsis," "The Yellow Violet," "Inscription for the Entrance to a Wood," "To a Waterfowl," "Green River," "A Winter Piece," "The Rivulet," "A Forest Hymn," "The Past," "To a Fringed Gentian," "The Death of the Flowers," and "The Battlefield," and you have Bryant's message; the rest of his work either echoes the notes already sounded in these, or represents uncharacteristic, and therefore transitory, moods.

Not less conspicuous than his excellences are Bryant's limitations. We may say of him that, like Wordsworth, he did not always overcome a tendency to emphasize the obvious, and that, like almost all contemplative poets, he sometimes made the didactic unnecessarily obtrusive. We have all heard parsons who, after finishing their sermon, sum it up in a valedictory prayer, with a hint as to its application, for the benefit of the Lord; equally superfluous, even for mortal readers, is the moral too often appended to a poem which is well able to convey its meaning without it. In this respect Bryant resembles most of our American poets, in whom didacticism has prevailed to an extent that will lessen their repute with posterity; for each generation manufactures more than enough of this commodity for its own consumption, and cannot be induced to try stale moralities left over from the fathers.

Bryant's self-control, the backbone of a character of high integrity, prevented him from indulging in emotions which, if they be not the substance of great poetry, are the color, the glow, which give great poetry its charm. He addresses the intellect; he has, if not heat, light; and he does not, as emotional poets sometimes do, play the intellect false or lead it astray.

In his versification he is compact and stately, though occasionally stiff. He came at the end of that metrical drought which lasted from Milton's death to Burns, when the instinct for writing musical iambics was lost, and, instead, men wrote in measured thuds, by rule. That phenomenon the psychologist should explain. How was it that a people lost, during a century and a half, its ear for metrical music, as if a violinist should suddenly prefer a tom-tom to a violin? Probably the exorbitant use of hymn and psalm singing, that came in with the Puritans, helped to degrade English poetry. The spirit which expelled emotion from worship, and destroyed whatever it could of the beauty of England's churches, had no understanding for metrical harmony. Any poor shred of morality, the tritest dogmatic platitude, if stretched thin, chopped into the required number of feet, rhymed, and packed into six or eight stanzas, with clumsy variations on the doxology at the end, made a hymn, for the edification of persons whose object was worship and not beauty. As a means to unction, mere doggerel, sung out of tune, would serve as well as anything.

At any rate, the taste for rigid iambics would naturally be acquired by Bryant at his church-going in childhood, and from the eighteenth century poets whom he read earliest. The beautiful variety of modulations which Coleridge, Shelley, Keats, and Tennyson have shown this verse — the historic metre of our race — to be susceptible of, lay beyond Bryant's range. His verse is either simple, almost colloquial, or dignified, as befits his theme; even in ornament he is sober. As he never surpassed the grandeur of conception of "Thanatopsis," so, I think, he did not afterward equal the splendid metrical sweep of certain passages in that wonderful poem.

And this fact points to another: Bryant is one of the few poets of genuine power whose poetic career shows no advance. The first arrow he drew from his quiver was the best, and with it he made his longest shot; many others he sent in the same direction, but they all fell behind the first. This accounts for the singleness and depth of the impression he has left; he stands for two or three elementals, and thereby keeps his force unscattered. He was not, indeed, wholly insensible to the romanticist stirrings of his time, as such effusions as " The Damsel of Peru," " The Arctic Lover," and " The Hunter's Serenade," bear witness. He wrote several pieces about Indians, — not the real red men, but those imaginary noble savages, possessors of all the primitive virtues, with whom our grandfathers peopled the American forests. He wrote strenuously in behalf of Greek emancipation and against slavery; but even here, though the subject lay very near his heart, he could not match the righteous vehemence of Whittier, or Lowell's alternate volleys of sarcasm and rebuke. Like Antaeus, Bryant ceased to be powerful when he did not tread his native earth.

We have thus surveyed his poetical product and genius, for to these first of all is due the celebration of his centennial, and we conclude that his contemporaries were right and that we are right in holding his work precious. But while it is through his poetry that Bryant survives, let us not forget the worth of his personality. For sixty years he was the dean of American letters. By his example he swept away the old foolish idea that unwillingness to pay bills, addiction to the bottle and women, and a preference for frowsy hair and dirty linen are necessary attributes of genius, especially of poetic genius. He disdained the proverbial backbiting and envy of authors. As the editor of a newspaper which for half a century had no superior in the country, he exercised an influence which cannot be computed. We who live under the *régime* of journalists who conceive it to be the mission of newspapers to deposit at every doorstep from eight to eighty pages of the moral and political garbage of the world every morning, — we may well magnify Bryant, whose long editorial career bore witness that being a journalist should not absolve a man from the common obligations of moral cleanliness, of veracity, of scandal-hating, of delicacy, of honor.

Finally, Bryant was a great citizen, — that last product which it is the business of our education and our political and social life to bring forth. In a monarchy the soldier is the type most highly prized; but in a democracy, if democratic forms shall long endure, citizens of the Bryant pattern, whose chief concern in public not less than in private life is to "make reason and the will of God prevail," must abound in constantly increasing numbers. Happy and grateful should we be that, in commemorating our earliest poet, we can discern no line of his which has not an upward tendency, no trait of his character unfit to be used in building a noble, strong, and righteous State.

www.ingramcontent.com/pod-product-compliance
Lightning Source LLC
Chambersburg PA
CBHW031859220426
43663CB00006B/692